Project Risk Management

Bruce T. Barkley

McGraw-Hill

New York Chicago San Francisco Lisbon London Madrid
Mexico City Milan New Delhi San Juan Seoul
Singapore Sydney Toronto

The McGraw-Hill Companies

Cataloging-in-Publication Data is on file with the Library of Congress.

8 9 0 IBT/IBT 0 1 0 9

ISBN 0-07-143691-X

The sponsoring editor for this book was Larry S. Hager and the production supervisor was Sherri Souffrance. It was set in Century Schoolbook by International Typesetting and Composition. The art director for the cover was Handel Low.

Printed and bound by IBT Global.

This book was printed on recycled, acid-free paper containing a minimum of 50% recycled, de-inked fiber.

To the over 3,000 hard working, adult graduate and undergraduate students at DeVry University/Keller Graduate School of Management—Atlanta, and at The University College, University of Maryland, who have provided me over the past 30 years with wonderful opportunities to learn from them—undoubtedly more than they learned from me.

Contents

About the Author

BRUCE T. BARKLEY has over 30 years of experience in program and project risk management in both industry and government. The coauthor of a successful book on project management, *Customer Driven Project Management: Building Quality into Project Processes*, Second Edition, Mr. Barkley has succeeded in making risk management clear and practical in a field that has become highly technical and quantified. Mr. Barkley has consulted in project risk management and serves as faculty member with DeVry University (Keller Graduate School of Management) in the Atlanta area. He is a graduate of Wittenberg University, University of Cincinnati, and the University of Southern California, has previously taught project management at the University of Maryland University College, and served as a senior executive in several federal agencies in Washington, D.C. He lives in Atlanta with his wife Cathy of 44 years.

Preface

This is not your father's risk management book—it is not your conventional treatment of project risk management. Rather than treating project risk as a narrow project and task-specific, "process" issue, risk is seen here as the outcome of bad project selection, bad business planning, and bad company-wide culture. Readers will experience a refreshing new perspective on project risk that centers risk management on:

- The business, enterprise-wide level
- Good business and project planning and management practice
- Building a healthy organizational culture that recognizes risk as the consequence of bad planning

Chapter 1 will offer new insights on building a risk management culture, while Chapter 2 deals with project selection using weighted models, financial return, and other risk information. Chapter 3 provides more useful tools, and Chapter 4 includes a broad interpretation of the current Project Management Institute PMBOK (Project Management Body of Knowledge) risk management section. Chapter 4 illustrates a good project management manual to offset risk while Chapter 6 illustrates a basic risk tool—the risk matrix table. Chapter 7 covers the use of Microsoft Project to calculate a "risk-based" project schedule. Chapter 8 revisits risk and quality management. Chapter 9 is another practical example of enterprise-wide, strategic risk planning in the Eastern case. Chapter 10 addresses how to document lessons learned in risk management.

Appendix A is a useful set of exercises and problems—with answers—on risk and cost management, Appendix B is an example of a risk-based schedule using MS Project, and Appendix C provides a checklist to demystify business and project risk management.

Acknowledgments

The author would like to acknowledge the following sources for this book:

- The Universal Avionics Systems Corporation, Instrument Division, for valuable experience in supporting and managing product development projects and processes, and writing program manuals and policy documents and conducting analyses in the program management office.
- The Alumax Aluminum Company (now Alcoa, Inc.), where the author was a project management and organizational development consultant, for valuable experience and case material in strategic planning and SWOT analysis in a manufacturing work setting.
- Keane, Inc., for valuable insights into how a successful, major software development and information company handles risk. Keane is one of the largest and most successful software services in North America, a company that has been at the forefront of project and risk management since the company was founded in 1965. Keane continues to offer such services and has trained thousands of client participants. The author used some of Keane's concepts in this book as examples of quality risk management.
- Adult students and faculty at DeVry University and Keller Graduate School, Atlanta, where the author serves as senior faculty member and curriculum manager for project management, for valuable stories, cases, and exercises in project and risk and cost management, which serve as the basis for material in the book. Special thanks to MBA student Jerel Hayes. His work on "Good Flight and Airlines" in chapter 2 and other student material have been used liberally in the book.

About This Book

Project risk management is an art, not a science. I have always been skeptical of scientific and overly quantitative answers to complex social, organizational, and project outcomes, especially when customers, products, and markets are involved. I think risk can be *stewarded* and managed by good planning and analysis, but in the end it is often the *gut feel* of a project manager that turns a project in the right direction and overcomes risk.

We tend to look for ways to control business and project outcomes that sometimes simply cannot be controlled. It is as if there is some underlying need to explain why things go "south" in a complex endeavor or project in systematic terms, and as if the world of human systems operated in a predictable and controllable way. We seek answers for all failures to *fix* them, yet we often do not know what factors were important. We assume that a *system* is in place and if the system fails we want to find out why it failed. When a business or project fails, we conclude that "somehow this failure could have been avoided if we had just studied and analyzed the risks a bit more, perhaps drilled a little deeper into the inherent impacts and probabilities."

Such an approach assumes that risks and failures operate in a predictable way, that the factors that lead to risk events and failure can always be identified, catalogued, and controlled, and that more analysis will uncover the *secret to the mystery.* The principle is that we should be able to identify what might happen, what the probabilities are, what the impacts are, and how to respond. It assumes that we can find attribution, that we can attribute failure to key events or circumstances. It is true that the root causes of project failure are rarely a mystery—they often have to do with business performance, market conditions, leadership bias, and lack of support. They rarely have to do with technological failure—the engineers will usually find a way—it's the organization that cannot stand success.

The problem is complicated by the variety of definitions among stakeholders of *failure and success.* One person's failure is another person's success. A project which overcomes technology risk can deliver within budget and schedule and be termed a success by the project team, but it is possible that this same deliverable cannot be manufactured, or that the customer is not happy with the outcome, or that the business itself fails for reasons that have nothing to do with the project.

The drive to *mystify* risk assumes that there is always one *true* risk involved in every factor, task, or project, and that to solve the risk mystery we have to go to extreme limits to identify and quantify that risk. This makes the subject more complicated than it needs to be—and assumes that is within our grasp to capture all the root causes of risk. Somehow if we can establish that the risk of failure of a team task to integrate an information system is 66 percent rather than 24 percent, we can make decisions based on an unreal confidence in science to predict things like the economy.

What is missing here is the fact that businesses and projects are human, not mechanical systems. Despite our increasing propensity to consider the study of organizational and project efforts in business to be a science rather than an art, human behavior is often unpredictable and counterintuitive. Despite our understanding of complex systems we cannot identify all the factors that contribute to risk and success even if we all agree on the definitions of these terms.

In addition, technical professionals and engineers have developed their own language and values, which sometimes complement but often conflict with good project risk management. The people and communication issues in engineering and product development are not unique, but they are accentuated by a working "axiom" of engineering project management—engineers communicate through channels and thought processes sometimes at odds with *cheaper, faster, better*. But they are inherently good risk managers. Engineers and technicians are often conflicted in a project management setting by time, cost, and organizational constraints that require them to take shortcuts to good engineering and risk management. They are challenged by risk and typically want to get it right, rather than getting it on time and at lowest cost. For instance, the measure of *mean time between failures* (MTBF) is often applied to electronic and technical equipment, and tests are designed to ensure that products perform under stress at the intended MTBF. MTBF is a risk indicator; the risk of failure is quantified by repeated tests and documentation. Thus a quantitative probability can be applied to its future performance. But in most circumstances MTBF is not applicable or suitable because user settings and environments cannot be controlled to really predict all the circumstances a product will experience. And the customer may not be interested—or will not pay for—a certain level of MTBF. But engineers typically would like to get MTBF down to zero if they can—an application of six sigma thinking—even at the cost of on-time delivery and budget.

Another complication in project risk management is the resistance to change in the project management or supplier team as well as in the customer's organization. Typically, a complex project and its outcomes trigger the need for organizational change, thus surfacing the resistance of those who do not see the value of change. For instance, a new electronic product produced through a product development project can alter the priorities of the customer's organization as the new product is phased into marketing and sales. The priority on this new product can upset an ongoing dynamic in the organization long supported by the *old, replaced* product. The risk here is that employees will resist change and undermine new product delivery unless the following factors are in place:

1. Top management support
2. Clear vision
3. Incentives to accept change
4. Incentives to take risks
5. Clear communication
6. No *walk the talk*
7. What Daryl R. Connor, founder and CEO of ODR, calls the Long View, a total, in-depth understanding of the effect the project will have on the organization

All this said, I am optimistic that there are useful tools to manage project risk and that these tools lie in core business and project planning and management processes. I believe that project risk can be *stewarded* but not always controlled through good planning and scheduling and critical thinking. Through the application of risk management tools outlined and illustrated in this book as part of the planning and control process—and separate from it—risk can be managed.

The book is designed for general reading in business, program, and project management, and for training and academic courses in project management and risk. The approach is to broaden and simplify the risk concept at the same time, offer useful tools and best practices, integrate risk into the strategic, business, and project planning and control processes, and to offer exercises and cases for learning purposes. The book shows how to apply the PMI Body of Knowledge on risk, but goes beyond it in many respects.

Figure 1 shows the *simple steps of risk management* that underlie the book, a sequence of topics that align with the chapters.

There is a simple logic in this book that provides the structure and sequence of chapters (Fig. 2). First, prepare the organization and its people for risk management as part of the business, then identify risks as part of business planning, integrate the process into project planning and control, then make it simple by using templates, and finally, learn what works.

Here is the logic in more detail.

- *First, prepare your organization or it won't work.* Risk management does not work unless everyone does it all the time and when management expects it to be done right.
- *Second, get started with good business plans.* Risk management does not start with projects; it starts with how the business is structured and planned.
- *Third, risk is not science; it is art.* Risk tools are simple planning aids and easy to apply. Don't overdo it and don't overquantify the process.

Prepare the organization: Build it into the culture
It starts with the business itself: Risk is embedded in the business
Use doable tools: Use project risk tools, they are easy
Demystify PMBOK: Learn PMBOK, but don't stop there
Use manuals to make risk policy: Make policy with risk manuals
Use easy forms: Forms make it easy
It is hard to do without MS project: Use PM software – it works
Products create risk: If you make a product, you have risk
Risk is customer-driven: Risk starts and ends with the customer
Risk makes strategy useful: Risk is the reason you plan strategy
Find out what works: Focus on what works, not what doesn't

Figure 1 The simple steps of risk management.

- *Fourth, the professional Project Management Institute Body of Knowledge on risk is helpful from a process view, but has limited applicability.* Make sure you know the *PMBOK*, but don't stop there.
- *Fifth, making risk policy in a simple manual.* Help project managers manage risk with a simple manual.
- *Sixth, see how others do it.* See the value of cases and exercises.
- *Seventh, use templates to make it easy.* Use forms already available.
- *Eighth, it's hard to do risk without MS Project.* You can do risk manually, but project management software makes it easy.
- *Ninth, you have to do risk in product development.* Whenever you produce products, you have risk.

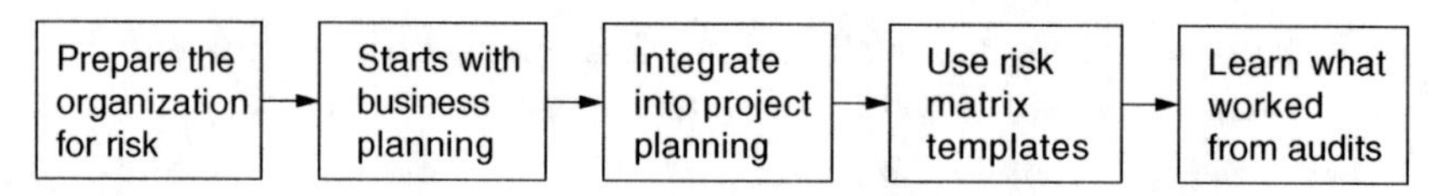

Figure 2 The logic of risk.

- *Tenth, risk is customer-driven.* You have to get in the shoes of the customer before you really understand risk.
- *Eleventh, risk makes strategic planning useful.* Risk is the only reason to do strategic planning; here is a case where risk is integrated with long-term planning, along with a presentation format.
- *Twelfth, after the fact, learn what worked.* Don't focus on mistakes; focus on what worked!

Introduction

The demystification of risk involves a whole new perspective on a business-wide process that has been looked at for many years as a separable, quantitative, and project specific exercise. The overkill in quantification comes from the attempt to replicate scientific, mathematical models of probability, but most projects do not need such rigor. The issue in project risk management is simple awareness of risks and intense management. As Fig. I.1 indicates, this book changes the paradigm for risk management while recognizing the value of current approaches.

Setting up for risk management means preparing the organization and not the project first. The issue is establishing the value of risk analysis as part of the normal project planning process.

Finding out where risks are is built into the work breakdown structure (WBS) and scheduling process; risk is an input to risk-based scheduling.

Dimensioning risk is qualitative, ranking and ordering, usually not quantitative.

Corrective action is not preparing separate task-based contingencies and kicking them in when necessary, but rather building them into the baseline schedule.

Postmortem audits by outsiders are rarely helpful because they are not accepted and do not reflect the insights of those who did the work; lessons learned meetings are far more helpful.

What Is Risk?

Risk, which is uncertainty that has been defined, is a simple concept, a way of thinking through and planning a program or project. There are many treatments of risk in the literature, but most tend to overdo the quantitative tools and understate the softer, more people-oriented issues in risk management. This book stays with the middle ground, touching all aspects of risk hopefully in a readable way.

The *demystification* of project risk involves some new assumptions about project planning and control.

First, risk has been narrowly treated in the context of projects and project tasks, but the sources of risk are more appropriately addressed at the business and industry level first. The prevailing notion about project risk management

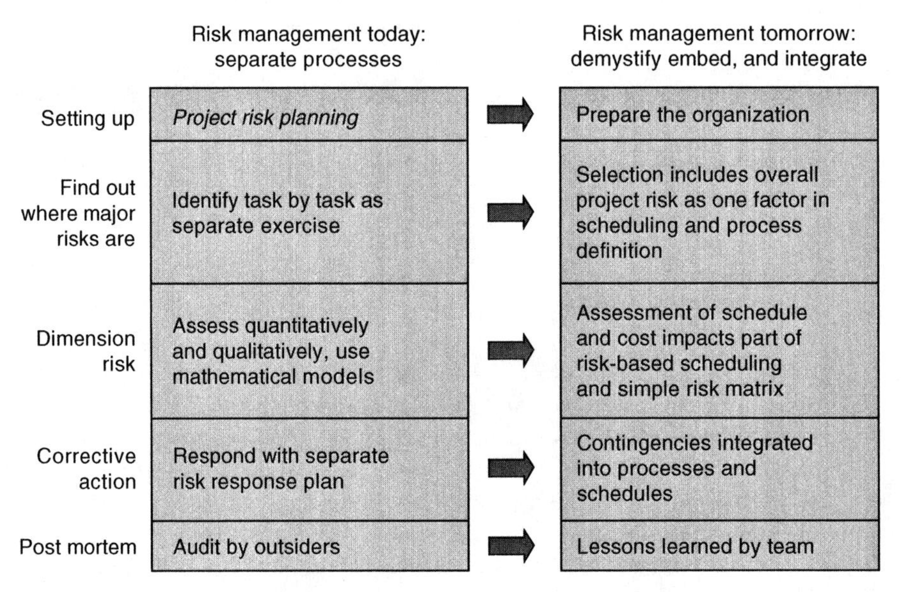

Figure I.1 Demystifying risk management: today and tomorrow.

has been the assumption that knowledge of internal, project-oriented planning and control issues was most important in forecasting and managing risks and costs. This assumption has driven the subject of project risk management in directions that focus on internal project tasks and risks. But business analysts increasingly find that emerging external business issues often have a much greater impact on the future of their organizations—and on project success—than *any* internal issues. Thus the roots of project risk lie in the forces acting on the company, and the customer, as a whole.

Second, and as a consequence of the first point, project risk cannot be separated from business planning, project selection, planning, and control. It is integral to these processes. Risk is the core planning challenge at the heart of business development and later, project management. The separation of risk management process from the rest of the broader business and project management paradigm is the wrong approach to the subject because it implies that somehow risk is largely internal to a project and therefore controlled by the project team. Since project risk is business risk, the whole business strategic planning, marketing, and risk analysis process is directly relevant to project risk. Risk applied to a business framework produces SWOT (strengths, weaknesses, opportunities, and threats) analysis and other outputs that support identification of project risks. These risks include competition, unanticipated technology change, market shifts, business finance, workforce issues, and changes in the customer base.

Third, risk management is largely a leadership and management challenge first, not fundamentally a quantitative process as portrayed in texts on the subject. Organizational culture drives the approach to risk. Risk is actually qualitative and intuitive and brings out the most creative juices of project process. It is risk that generates the passion of business achievement; to overcome risk is to overcome a competitive challenge and create opportunity. Overcoming risk equals business success.

This book addresses *the process* of identifying, analyzing, and responding to business and project risk in order to minimize the consequences of adverse risk-based events. The PMI *PMBOK* processes of risk planning, identification, quantification, response planning, and control are covered, as well as risk factors, contract types, assessment techniques, tools to quantify risk, procedures to reduce threats to project objectives, and contingency.

Risk definitions

The Software Engineering Institute defines risk management as "A successful risk management practice is one in which risks are continuously identified and analyzed for relative importance. Risks are mitigated, tracked, and controlled to effectively use program resources. Problems are prevented before they occur and personnel consciously focus on what could affect product quality and schedules."

You can see that this definition is fairly broad and describes a process that goes through the whole project life cycle. The definition addresses the way project team members think and act in the planning and organizing process.

There are five principles underlying the definition of risk in this book:

1. Risk is any uncertainty in a project plan that you can potentially control, or at least track. This means that there are many risks in any project. The trick is to identify the most critical risks—the ones that could make or break your project—and control them. Overcoming a risk—that is being able to complete a project or project task despite the risk—creates opportunity. The other side of risk is opportunity—if a business is *better, faster, and cheaper* in producing its products and addressing customer needs and reducing the risk in the process at the same time then the payoff opportunity is market share and business growth.
2. Risk is integral to the business and the project planning process; therefore don't think of risk as something different or separate from management. Risk is why you do business and plan projects— if there were no risk, there wouldn't be a project. And addressing risk simply means that you are always looking around you to find things that can go wrong in defining and scheduling work.
3. Focus only on the high-risk, resource-consuming tasks because you can't focus on all of them all the time. Assessing risk is a question of rank-ordering risks and keeping your eye on them. While we will exercise some quantitative tools in this course, such as probability analysis, these tools have very selective applications when you have a very complex project and you have background data on technical probabilities.

4. Monitoring risk is a question of identifying key risk milestones or points in the project schedule where risk decisions need to be made. These milestones would mark whether a piece of equipment worked, or a key resource was available, or a key technology in a new product worked as designed.
5. Planning a response to risk involves understanding the project and impacts of various corrective actions midstream. You create risk scenarios and schedule impacts. An "expected" scenario is the best guess at what actually will happen, a "pessimistic" scenario is the worst case, and an optimistic scenario is the "best case."

Risk, Process, and the Myth of Control

There is a natural tendency to study risks as separate problems, to see risk as one shot points-in-time when risks are systematically identified, assessed qualitatively and quantitatively using sophisticated mathematical models, and controlled through conceived contingency plans. But real-world experience teaches us that risk is, in truth, an inseparable aspect of the whole project life cycle and its daily irrationality and interpersonal dynamic. In a way, risk events are a result of bad planning. In that sense, risk can be seen as a continuous series of individual and collective decisions in planning and managing a project. The process is not mystical and quantitative; it is organic and intuitive. You head off some risks while creating others—you mitigate a technology risk with information on impacts, and address the potential disease, but there may be risk in the cure as well. Some risks occur despite mitigation, while others do not, despite being ignored. Sometimes neglected risks never happen.

Many decisions add up to a successful management of risk. The tyranny of small decisions adds up to success or failure. Thus risk is part of the planning cycle—planning is designed to reduce risk, but in fact the role of planning is to see risk coming and to address it *in the plan.*

Risk and quality are as cost and benefit. Quality may be defined as the extent to which the road taken (the process) conforms to the proven road already taken successfully. To know your process and follow it consistently is to produce a quality product. This suggests that if you know the correct process, you can produce quality at minimum risk and cost, simply because you know how the process works.

But if the process and product are new, the cost of quality increases as you incur costs of waste, redundancy, and inspection/appraisal. Risks unattended create costs because they imply repetition, error, and delay. But these costs are not simply costs of delay and schedule; they are costs of actions, contingencies.

For example, in the design and production of a complex, electronic, digitized avionics instrument, there is an inherent risk at every step in the process:

1. *Customer requirement.* There might not be a "customer" per se, or if there is, the customer has no idea what is needed. Thus customer requirement itself

is a risk and that is why we spend time trying to identify requirements. Requirements analysis is a risk reduction tool. Each time we assume that we have the requirement down and choose not to test it out with the customer—or each time we do test it out but the customer falsely assures us that the specification is correct—we add incrementally to the totality of the customer requirement risk. Checking with the customer is not the simple solution, as the customer changes expectations; risks that they may be unrealizable change, sometimes for the better. In other words, as the customer is educated on risk, the customer is liable to make decisions, which lessen his or her risk and sometimes the risk the project itself faces.

2. *Concept.* The concept might be flawed because it is not feasible. Innovation in project concept leads to creative vision of what is possible, but it might not be possible or even desirable in the eyes of the customer or the key stakeholder. Thus there is inherent risk in designing a concept but there is also opportunity. The more options are presented to the customer in the concept stage, the more the probability that one or another will delight the customer, at least in principle. Thus innovation applied in the concept stage is a risk mitigation step in the sense that it is in this stage where ideas and visions can be addressed without substantial cost.
3. *Design.* The design might not meet the requirements, or the design might be feasible in production and assembly. Design, putting a concept into a drawing or rendering, involves a myriad of steps that are inherently risky; from misstating tolerances to choosing unavailable components, to designing correctly to an inaccurate requirement or specification.
4. *Prototype.* The development of the prototype may not be aligned with the requirements, so testing the prototype does not assure success either in conformance to specifications or customer satisfaction. In building a physical model of the product and testing it, there is inherent risk that the prototype is not representative of the (unexpressed) expectations of the customer, or that the prototype is not exactly like the product that is to be manufactured. Or the costs of the prototype may not be accurate so that when the time to produce it arrives, the company cannot afford it at the price used in justifying it.
5. *Production.* The product process might not be consistent with the time and resources needed to put the product together and produce it in volume.

Risk/Benefit

It is important to see risk as a tradeoff with benefits, opportunities, and payoffs. In other words, risk is the reason for investment—to seek out profitability by reducing uncertainty and gaining benefits in terms of customer value and profitability. The following matrix (Fig. I.2) illustrates the tradeoffs involved in categorizing and selecting projects for a business portfolio.

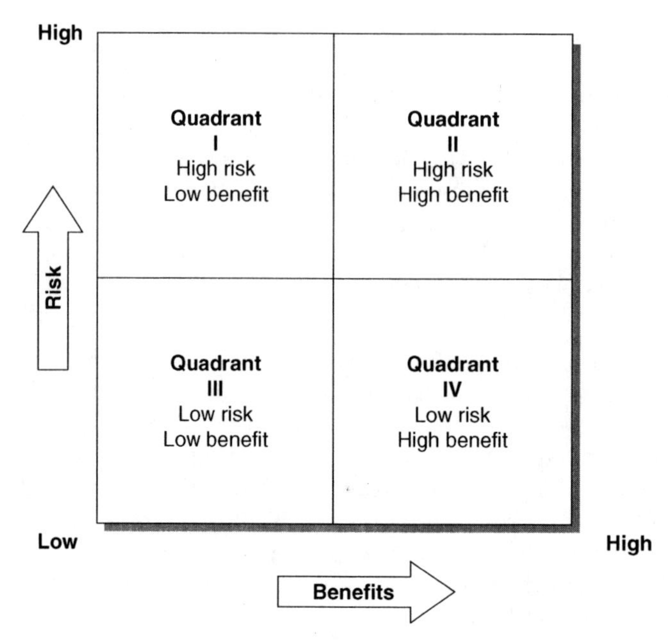

Figure I.2 Risk/benefit template.

Quadrant I: High risk, low benefit

Projects which fall into this quadrant are not worth doing simply because there is great uncertainty about outcomes and little foreseeable payoff. Of course, companies typically support a limited number of exploratory R&D projects, some of which can move from this quadrant to the next when unexpected payoffs are uncovered.

Quadrant II: High risk, high benefit

Projects here are major investments with high risks of failure, but with outcomes that could substantially improve market share, company growth, and profitability. A good example of such a project would be an investment project in the field of space exploration.

Quadrant III: Low risk, low benefit

Projects in this quadrant are not worth doing simply because there is no foreseeable payoff, even though the cost or risk involved is minimal. An example would be a superficial landscaping improvement to a plant location when the permanence and viability of the plant itself is in question.

Quadrant IV: Low risk, high benefit

Here the projects are very attractive because for minimal risk there is a potential high benefit. An example would be installation of a proven technology in manufacturing that promises to double productivity of a current plant or facility.

A Way of Thinking

Risk is a way of thinking first, a balanced worldview that looks critically at business, program, and project decisions in terms of both sides of the question. In this book, we will explore various ways of *embedding* this way of thinking into the organization so that risk and benefit tradeoffs are part of all key decisions.

Demystifying Business and Project Risk Management

Risk management involves a whole set of activities that are embedded into the project planning process. Appendix D summarizes the kinds of risk checklist actions that are covered in this book.

Chapter

1

Preparing the Organization: Building a Risk Management Culture

Building a culture of risk management is primarily a process of developing people in your organization who think and plan projects effectively, and who are supported by company systems that encourage them to think and plan effectively. That involves looking constantly at what could go wrong and knowing the difference between theoretical risk and practical risk. Theoretical risk is risk that *could* happen; practical risk is risk that is *likely* to happen. Experience helps to differentiate the two.

Prepare the Organization

If the organization does not address risk in *the way work is done*, risk management will fail. Defining culture as the way work is done in the organization, if risk is integrated in the way work is done (e.g., project plans incorporate a risk matrix as defined later in this book) risk planning becomes an expected part of planning. If risk is given lip service but not backed up, then risk management will be superficial and ineffective (Fig. 1.1).

The best way to illustrate risk is to tell a little story about what happens when risk is not in the organization's culture. See if you can relate to the following story.

> "I got the customer's approval to do the Schneider program," Lakeisha told Bill. Lakeisha and Bill were project managers at Project Associates, Inc., a software and information technology company. "I've got 4 months to deliver the program to Schneider, including a new hardware platform, software code and documents, and a training manual. I think it's going to be a blast—the biggest issue to me is the software. The hardware is a no-brainer."

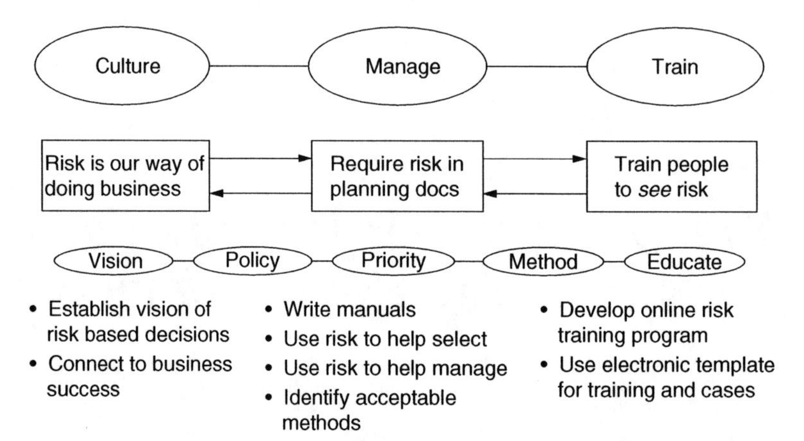

Figure 1.1 The process of preparing the organization.

Lakeisha had delivered her last project, the Mires program, well ahead of schedule. Bill, on the other hand, had not done well in his last project and was late and over budget. Lakeisha was eager to show that her last performance was not a fluke and that she knew what she was doing. She always harbored little lack of confidence under pressure, but always came through. There was a subtle competition going on between the two, but they worked well together.

"I wouldn't get too excited just yet," Bill told her. "You know I was going to do that program, but I got sidetracked and the boss gave it to you. I've seen the specification for the program, and there are a lot of risks. I think you've got at least 6 to 7 months of work with the current team to produce the deliverable as I see it. And that is if you don't have any problems with platforms, software, people, our old testing equipment, and good old unreliable software systems, our contractor. Are you going to go with the same team you used last time? Planning any risk assessment and contingency—you know, those scenario things?"

"Yes and no. I am going to use the same team, I think, but I don't have time for the risk stuff this time. It isn't a required part of the project plan, especially if you have been there before as I have. Been there, done that. I will use a schedule from our last project for Smothers, which was a good run and we kicked butt. I have looked at the risks and the big ones are in the software graphics package and online training package. I've got a plan to shrink the schedule to 4 months by crashing some stuff and outsourcing. I read an article on outsourcing last week and the story included a great company that does just what I want them to do—or almost. That is going to cut my schedule by 2 months. This project has to be on the fast track from the word go. I am going to write a cost plus contract with them because procurement says that is the template they like to work with."

"Well, I hope you know what you're doing," Bill said to Lakeisha as they crossed in the corridor. "I have seen a lot of people get burned by contractors and you know that the hardware for this deliverable is new and will require some long lead times on parts."

"Cool it Bill. I've picked a reputable contractor," Lakeisha said. "I checked his references—actually I just talked to a friend who works there—and I am sure that the contractor will do a great job and will accept most of the risk. I told him I would make progress payments on the cost plus contract only if he was on schedule and he agreed. He said he would put more people on the job if it turned out to be more complicated than he thought it was. I will just keep an eye on him. I understand that his is a risky business and some risks are inescapable. But what an opportunity to show what we can do—think of the plus side of this thing if it goes! When I have this much to do in such a short time I'm not going to waste the team's time on risk games and risk matrices. I know what I want and I know what the risks are."

Bill thought she should be more careful, but liked her spirit. He and Lakeisha had been over this ground before. He had learned some lessons on risk in his previous project, which failed because he was hit by an unanticipated shortage of key technical people and a surprise glitch in getting parts for the hardware platform. He had been reading about a new theory of "constraints," which focused on resource and equipment availability risks and not just critical path issues. But he had learned not to argue with Lakeisha when she had already decided what she was going to do.

Lakeisha met with the contractor and gave him the specification before her procurement office could get the scope out and signed, but they didn't have a problem with that and she didn't have time to go through formal signatures anyway. It turned out that the contractor, Mag Company, was headquartered in New Delhi, India, but had a local office. Their procurement people had said the specification made sense and that they would get on it right away—they too felt that it could go ahead without a signed contract. They needed the work.

Six weeks later, Lakeisha called the local Mag Company project manager, Abdur Manat, to check on progress. "Everything is going great," he said. "But we have been working on a high-priority project for another company who came in just last week with a heavy job due yesterday, of course. So I have not made as much progress as I had hoped." Lakeisha responded, "Maybe I ought to have a schedule from you, particularly on the tough pieces of the work you see as potential problems." Abdur replied, "But I still have 3 1/2 months to do 2 months of work, so I don't see any problems. Did you send me those specs on the hardware? By the way, did you say that the customer wants online training—what's that?"

Lakeisha hesitated but disguised her concern in a positive expression. "That sounds fine," she responded. "Let me know if you need anything. I'll be back in touch in another 6 weeks, and then we can talk about integration."

For the next several weeks, Lakeisha spent most of her time trying to put together the specifications for an online training program for the contractor, a task she hadn't anticipated. She also inquired on the lead times for the new hardware, but the design people were busy on other work.

Six week later, Lakeisha called Abdur to check on progress. "The last project took me longer than I expected," Abdur said. "I've gotten into the graphics work and looked at your hardware requirements, and I've been working like crazy, but now that I have taken a closer look at it, I think there's at least a good 3 months of work on this job, particularly on that online training stuff you sent me—I will have to sub that out."

Lakeisha almost choked on that one. Her stomach told her she was in trouble and she murmured to herself. That would make the total development time for the Schneider project 6 months instead of 4 months. "Three months!" she said, "you have to be kidding. I need the software code in 2 weeks to begin integration. You were supposed to be done by now! I am not paying your last invoice."

Abdur responded, "OK, but you already paid our last invoice—your accounting office has been very efficient. We just got the check, along with a nice holiday greeting." Lakeisha knew this was not her day.

"I am truly sorry," Abdur said, "but this isn't my fault. There's more work here than we could have ever done and more than you estimated in your schedule. We found that the software code doesn't work in your hardware and we haven't been able to figure out why. And your team people aren't available to talk to. I will finish it as fast as I can."

It turns out that Abdur delivered the software in 3 months, but the project took another month after that because of integration problems with the in-house team's code. In the end, the Schneider program took over 7 months rather than the 4-month estimate. Lakeisha concluded that Bill and the company had "sandbagged her" by palming off a bad project that he was not able to handle and that had inherent big risks.

What is missing in this organization is a risk-based business culture. Lakeisha was treating the project in a careless and superficial way. She acts as if she is *the only* agent of project success. Her company has isolated her in a narrow project manager role without support systems, incentive, and training. A company that lets a project manager perform that way is a company that does not understand its own culture.

Risk: The Organizational Culture Issue

While risk is traditionally seen as an analytic activity (identifying and assessing risks in the project task structure, and applying decision trees, sensitivity analysis, and fine-tuned probabilities) the essence of risk management is the way your organization treats risk and the way you and your team think about the project. The challenge for the organization is teaching and training project leaders and team members to think in terms of risk and to internalize the risk management process into their daily work. They are the front line of risk management. The assumption behind this approach is that risk management is "something I want my people to do in the normal course of their work," not something I want a specialist to do later in the project as a separate audit exercise. Risk is a way of visualizing the project and its successful outcomes and *seeing* potential pitfalls. You can't see risks if you are not looking for them.

So the successful management of risk is usually the product of a successful organization that has instilled into its people the importance of careful planning. Careful planning involves a core competence—the capacity to dimension uncertainty and risk, to integrate risk identification and assessment into program and project planning, and to build and sustain a support system for risk management that provides essential information when it is needed. But how does an organization build risk into its daily work, and how do executives use their leadership and institutional leverage to further good risk management?

A Culture of Risk Management Competence

The successful risk management organization has five basic competencies:

- Active training and development in risk planning and management
- Strong linkage between corporate planning and project planning, particularly between business analysis of threats and opportunities, and analysis of project risk
- Deep project experience in its industry
- Capacity to document project experience and "learn" as an organization
- A workforce of strong functional managers who address product quality as a risk reduction issue

Link Corporate and Project Planning

Strong ties between corporate strategic planning, including market analysis, and project planning ensure that the business "sees" its technological risks early in its business planning and is able to anticipate and dimension the risks it will face in designing and implementing projects that carry out its strategies. For instance, a telecommunications firm that performs SWOT (strengths, weaknesses, opportunities, threats) analysis in its field may uncover a potential threat in unanticipated breakthroughs in telecommunications cable technology. Addressing contingencies at the corporate level to address these potential breakthroughs (opportunities created from analysis of threats) helps the business support its selected projects that involve such new cable systems.

Training and Development in Risk

Training and development programs that address risk identification, assessment, and response can help build professional competence in handling risk issues in real projects. Such training would include a curriculum in:

- Building a WBS (work breakdown structure)
- Identifying risks in the WBS
- Producing a risk matrix

Project Experience

A company that "sticks-to-the-knitting," as Tom Peters called it in *Search for Excellence*, is in a better position to recognize and offset risk simply because its workforce is likely to have a better handle on the technology and process risks inherent in its core business. Whenever a business departs fundamentally from its core competency areas, it stands to experience unanticipated problems, which develop into high-impact and high-severity risks.

Learning Organization

A learning organization, as Peter Senge describes it, is an organization that does not reinvent the wheel each time it plans and implements a project. This means that lessons learned from real project experiences are incorporated in documentation and embedded in training programs so that project managers learn from past experiences. Communication is open in such organizations, leading to a process by which project experiences are "handed" down to next generation project teams.

Strong Functional Managers Address Quality

The existence of strong functional management ensures that the basic functional competency of the company in areas such as engineering or system development is backed up by technology leaders in the field. Key processes like product development are documented and product components controlled through with disciplined configuration systems. This means that the risks of product quality failures that result from product component variation are minimized in methodologies such as six sigma simply because the company can replicate products and prototypes repeatedly for manufacturing and production without variation.

Building the Culture

Organization culture can be defined as the "prevailing standard for what is acceptable in work systems, work performance, and work setting." A *risk management culture* can be defined as the "prevailing standard for how risk is handled." An organization with a strong risk management culture has policies and procedures that *require* its workforce to go through disciplined risk planning, identification, assessment, and risk response project phasing.

A mature organization does not treat risk management as a separate process, but rather "embeds" the risk process into the whole project planning and control process. Risk is an integral part of the thinking of its key people. In the same way that the quality movement matures to the point that quality assurance and statistical process control processes become institutionalized into the company rubric, risk assessment tools and response mechanisms become an indistinguishable part of a company mosaic in a mature organization.

Sustaining the culture of risk management is considered a major function of corporate leadership in the *risk-planning* phase. Although most organizations do not enter the risk-planning phase as a distinct step in the project planning process, best practice addresses potentially high-risk tasks, assigns probability implicitly to the process, and develops optional contingencies that may or may not be documented in a formal risk matrix. This is typically not a mysterious, mathematical process, but rather an open, communicative process in which key project stakeholders, team members, and the customer talk about uncertainty and identify key "go or no go" decision points. They often know where the key risks are in the project process because the project itself is grounded in addressing a risk that the customer is facing.

Keane's Risk Process

A good example of a strong risk management culture is found at the Keane Company.

Keane connects and integrates risk with cost and schedule estimating, e.g., identifying project risks and determining actions to minimize the impact on the project and to improve project estimates. In other words, Keane thinks in terms of risk as a guide for cost estimating, scheduling, and defining mitigation actions.

The process starts with an estimating process that takes much of the guesswork out of estimating. Keane has established a set of guidelines, techniques, and practices to pin down estimates and to ensure that customers and stakeholders clearly understand associated risks. Keane emphasizes communication on the relationship between a given project estimate (project schedule and cost estimate) and how the estimate has handled risks and risk mitigation. Their experience is that project success does not depend as much on completely mitigating risk as on communicating risk up front so that stakeholders can make judgments and decisions along with the project team as things happen.

In building the culture for risk management, Keane warns its people about the hazards of estimating:

- Making sure they know the difference between negotiating and estimating. Estimating is the calculation of schedule and cost given the tasks at hand; negotiation is working out differences between the estimate and a customer or client schedule and cost.
- Understanding the variations in technical skill in how those variations can impact estimates.
- Being objective about your own work.
- Adjusting to the lack of an estimating database.
- Being too precise before it is needed, understanding the timing for order-of-magnitude, ballpark estimates, versus the need for more detailed budget and definitive estimates.
- Understanding the limitations of work measurement.
- Looking at untracked overtime in building estimates from past work.

Keane advises its people to ask the question "who is at risk?" before you ask the question, "what is at risk?" This is because the issue of risk is framed by those who are affected by it, not by some arbitrary quantitative formula. Different project stakeholders have different perspectives on risk and estimates, and indeed their perspectives change during the life of the project. It is best that risk assessment be guided by those who will suffer the consequences of risk and who will bear some or all of the cost of risk mitigation.

The role of the project planner/manager during this process is to inform the process with parametric data. Keane has found that in many cases the person asking for the estimate is more at risk than other stakeholders, or the project manager, really understand. This is because the person asking is going to use the estimate to make business critical decisions. For instance, if a client for a new information system is facing the possibility that a new system cannot handle the estimated user load on it projected for peak periods, then that client must make a decision either to limit the user universe or upgrade the system. So the estimate of risk is key to the client decision process and will affect client success.

Keane integrates cost and risk to better understand how risk effects project schedules. By training its people to identify risks from broader business and industry data and to schedule risk planning and management activity into the project baseline schedule, the company delivers an important message to its people.

Addressing Risk with Scenarios

Keane is a good example of a *projectized* company that uses risk scenarios to get its project teams to anticipate risks in the planning process. It encourages the development of issue or scenario statements that pose potential problems—variations from the plan—in a project and generate queries about the issue. For instance, Keane might encourage a project manager developing a new project information system to build the following question into an early project review session: What challenges does this new system create for the customer and what is the likelihood of these challenges becoming project "show stoppers," what case we do about it now?

Performance Incentives

Any organization building a risk-based culture must provide incentives for integrating risk into the project planning and control process. The incentive for handling risk is top management support and resources. Top management support comes when project management identifies and anticipates business risks that save the company time and money. Project managers who manage risks effectively are likely to be more successful in acquiring additional resources because they tend to have backup and contingency plans ready when risks occur.

Taking Risks: The Risk of "Blinders"

One of the major risks in any project is the tendency of its key project decision makers, especially the project manager, to overestimate what they know and underestimate what they don't know. The risk is that key people will "take risks" but not manage risk. This means that the beginning of good risk management is the capacity to know what the organization and its people can do and what they cannot do.

The field of organizational behavior contributes a tool called the *Johari Window* that is helpful in analyzing personal tendencies of project managers to take risks rather than manage them (Fig. 1.2).

The Johari Window, named after the first names of its inventors, Joseph Luft and Harry Ingham, is one of the most useful models describing the process of human interaction and behavior. A four-paned "window," as illustrated below, divides personal awareness into four different types, as represented by its four quadrants—open, hidden, blind, and unknown. The lines dividing the four panes are like window shades, which can move as an interaction progresses.

A typical project manager might go through the following thinking process personally to test what he or she knows:

1. The "open" quadrant represents both things that I know about myself and that others know about me. For example, I know my name and so do you, and

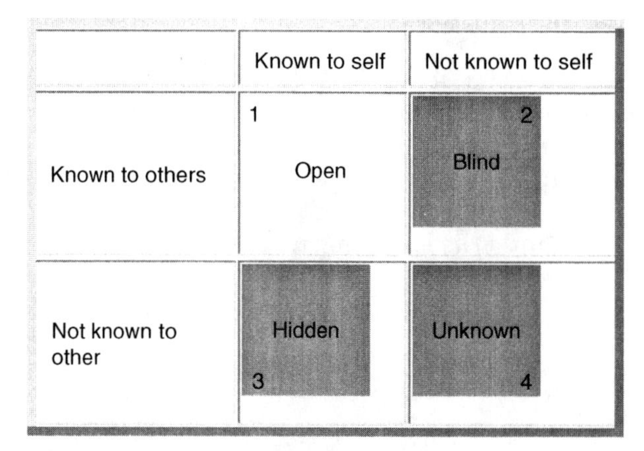

Figure 1.2 The Johari Window.

you know some of my interests. The knowledge that the window represents can include not only factual information, but my feelings, motives, behaviors, wants, needs, and desires. Indeed, any information describing who I am.

The risk here is that what is open to some coworkers may not be open to the customer or a project sponsor. So it is important that a project manager get to know the customer and key sponsors or stakeholders as people. The focus is customer expectations; if there is an open process on expectations then the chances of managing risk are high.

2. The "blind" quadrant represents things that you know about me but I am unaware of. So, for example, we could be eating at a restaurant, and I may have unknowingly gotten some food on my face. This information is in my blind quadrant because you can see it, but I cannot. If you now tell me that I have something on my face, then the window shade moves to the right, enlarging the open quadrant's area.

   The risk factor here is that there may be variables that a competitor or customer knows about the organization that the project manager may not know. For instance, a current supplier to the project may have failed in delivery of a similar component to a competitor, but the project manager is unaware of the situation.

3. The "hidden" quadrant represents things that I know about myself and you do not know. So for example, I have neither told you nor mentioned anywhere on my website, what one of my favorite ice cream flavors is. This information is in my "hidden" quadrant.

   The risk here is that a project team member may not be entirely open in divulging important information about their expertise and experience.

4. The "unknown" quadrant represents things that neither I know about myself, nor you know about me. For example, I may disclose a dream that I had, and

as we both attempt to understand its significance, a new awareness may emerge, known to neither of us before the conversation took place.

The risk here is that there are factors at work that are unanticipated both by the project manager and the customer.

Personal, Project, and Organizational Risk

There is something very personal about the issue of risk. In many companies, taking risks is rewarded in principle, but failure in taking risks has its implications despite the company rhetoric. What the company is really saying is, "Go ahead and take risks, but take them only if you think you can succeed and produce value for the customer and the company. We will support you with data and information. Don't take risks frivolously."

For the business and project professional, risk is first a personal issue because project risk is directly associated with personal risk. If a project manager fails to see and control risk, that project manager faces the prospect of being associated with a failed project. So the way a project team faces risk has implications for each team member personally—and for the team dynamics involved in a given project.

The way the company protects its employees and officers from risk is key as well. If the company is positioned to absorb the cost of failure then the program or project manager is more likely to take the risk. Thus the propensity to accept risk and manage it successfully is partly a function of organizational support—if my company supports me, I will address risk and make the best decisions I can, but I will want to let my top management know the risks as I see them so that if the risk is not successfully controlled, it will have been a company-wide decision, not a personal one.

In sum, the model is this—the organization must position itself for risk and must empower and enable its business and project people to address and take risks, but there must be an open, organization-wide process for addressing and absorbing risk. If these conditions don't exist, the project manager is not "incentivized" to address risk and will avoid risk, often at the expense of opportunity.

Chapter

2

The Business Risk Framework

It is important to see risk as a business-wide challenge. After all, business enterprise itself is a risk and that is what makes success and payoff satisfying to the business entrepreneur. Project risk is simply a microcosm of the overall business challenge and the fate of every project lies first in the capacity of the parent company to create conditions for success. As we have said, project risk starts with the business itself, its market position and business viability, its partnerships and vendor relationships, and the economic risks the business itself faces, as well as customer and client risks.

The author learned a valuable lesson in project risk management working with a leading electronic avionics product company—a product development and manufacturing company that used project management systems at the division level but had no project management systems in the corporate "head-shed."

As the support project management office, my role was to assure that there was support to the project managers and a clear project management policy and process and that standards were enforced, as well as serving as an assistant project manager in several functional areas such as procurement and cost capture. The work was in a regional facility in the East, while corporate was in the West run by a single owner and a small corporate staff largely without corporate program management competence.

This major producer of avionics equipment had several product development projects going on in the engineering plant involving mechanical, software, and electrical engineers, supported by a procurement and acquisition and accounting staff, and an HR office. The product involved embedded software and regulatory requirements for avionics equipment, and much of the process was testing and retesting prototypes against standards. In addition to the work underway, the regional facility had been *awarded* (by the owner's decision) a system program from another region that had not been successful with a system upgrade. New staff were hired to staff the project out, with the blessing of corporate. At the same time it was authorizing this hiring process at our facility, corporate was experiencing downturns in sales and marketing efforts and consolidating facilities to reduce costs. But the owner made the decision and we implemented it.

Later the same year the company had to conduct a major downsizing, cutting many of those same staff hired to run the new program, plus other valuable and high performing engineers. The reason it had to downsize was that the corporate investment source was unwilling to forward additional funding until the company cut costs.

We learned that project risk cannot be separated from business risk in general, and that the effectiveness with which a company identifies broad threats and risks in its business planning will establish the conditions for successful management of project risks.

Project risk is inherently business risk and cannot be disassociated with the overall risks and threats faced by the business as a whole. Thus project risk management must start at the perimeter of the business and its relationship to its market environment. As the business identifies its threats, competitors, and risks, it provides the basic wherewithal to identify project risks. There is an inextricable linkage here between the threats a company faces in technology, or labor availability, or product development, and the threats a project team faces in producing a deliverable designed to implement the business plan and strategy.

The lesson is this: risk is a *vertical* process, not just *horizontal*, that is, risk happens up and down the organization at the same time it happens in project planning management processes over time. Risk is multidimensional and multiscaled syndrome that can affect you without warning if you are not *in the inside*. And risk does not often come in recognizable clothes, but rather sneaks up on you through the side door. Very often the key determinant risk is out of your control as a program manager, or even as a general manager, because risk stems from central leadership more than it does through project processes.

The "risk as part of the business framework" concept is diagrammed in Fig. 2.1.

The challenge, given various project descriptions in a wide variety of fields, is to integrate broad business strategic planning and analysis of threats and risks with project risk management.

Knowing the business you are in helps anticipate risk inherent in the business. If I am the project manager of a software development project in a software firm, I know from past experience and good corporate strategic planning that one of my risks—as a business—is going to be the "integration and debug" process, the time and effort to make sure a new computer program or code works on the user's hardware platform. I can almost bet that this task will be one of my risks in any such project. I can differentiate that very distinct risk from a general uncertainty about whether there will be any demand for the product once it is ready. I can "work" the risk in my project, the uncertainty about market demand can also be worked; but I can't control it.

The risk in this software development process can be identified, defined, and ranked—a process we call *qualitative risk assessment*. If I wanted to quantify the

Figure 2.1 Business framework for risk.

probability that a debug problem will actually occur and create a major schedule slippage and/or quality issue, I would do a quantitative assessment of probability. That process might result my estimating that there is an 80 percent probability of a debug problem not getting fixed within my estimated and scheduled "most likely" task duration of 3 weeks. Then I might identify two alternatives—a worst case and a best case—based on various assumptions and plug them into my schedule using the "PERT" tool.

Thus the reason that project risk starts with the business itself is that any project that comes from the business pipeline, or portfolio of projects, is typically aligned with the business competencies and capacity. The business leadership has chosen a project because it believes the payoff of a project is worth the investment in overcoming the risks inherent in doing the project. The product or service to be produced by the project is key to the success of the business in its industry niche (Fig. 2.2).

Portfolio Management

The following is a case in portfolio management by Jerel Hayes.

Let's say that a new business, Good Flight Airlines, is a new consumer airline in its formative stages. It is being organized to take advantage of a specific gap in the low-cost international travel market. The gap in the availability of low-cost international service in and out of DeKalb-Peachtree Airport (PDK), Atlanta, Georgia, coupled with the demand for passenger travel on selected routes from PDK indicates that a new entrant airline could be expected to capture a significant portion of current air travel. The airline will initially provide scheduled service to destinations in Mexico. Initial focus cities will be Monterrey and Mexico City. Initially two Canadair CRJ 700 regional jets will be leased from International Lease Finance Corporation (ILFC) or GE Capital Aviation Services.

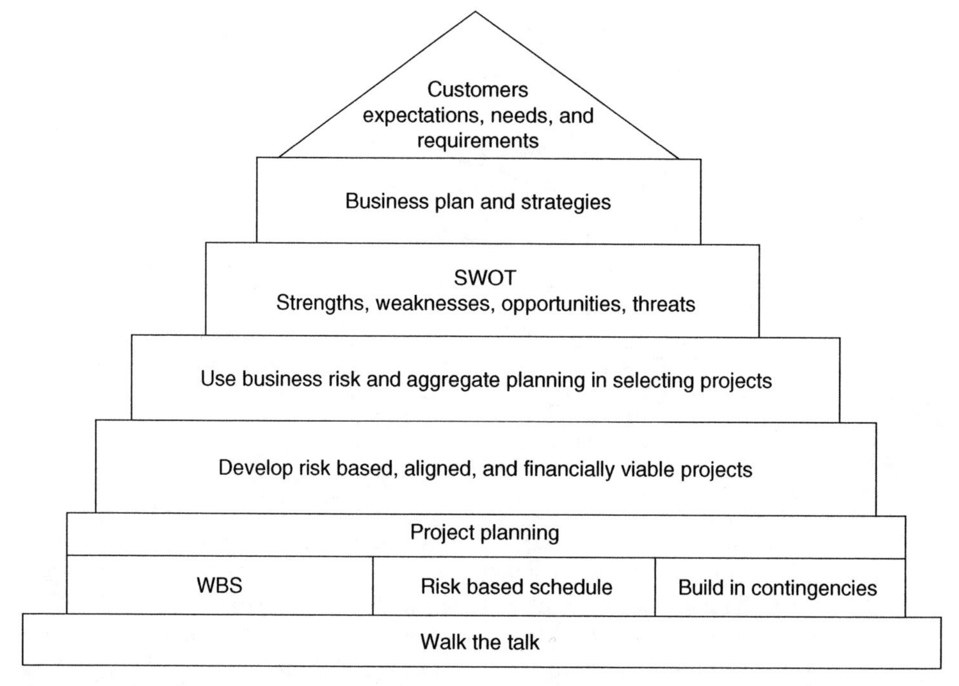

Figure 2.2 Business framework pyramid.

Organization

The airline will hire an experienced management team, full-time pilots, mechanics, flight attendants, and baggage handlers. Reservation agents will be outsourced. Initial plan includes leasing terminal space at PDK.

Strategic Statement

Good Flight Airlines is to hire innovative people dedicated to delivering the best flying experience to smart travelers, every day. Good Flight will be the first low-cost international regional airline in the United States and will strive for profitability in 2 years. Good Flight will open up new markets at smaller regional airports in proximity to larger international airports in cities with a high concentration of people of Hispanic descent.

One- to Five-Year Strategic Objectives

- *Objective 1:* To obtain required DOT and FAA certifications on or before June 25, 2004
- *Objective 2:* To commence revenue service
- *Objective 3:* To raise sufficient *seed* and *bridge* capital in a timely fashion to financially enable these objectives

- *Objective 4:* To commence operations with two Canadair CRJ 700 regional jet aircraft in month 1, four by the end of month 4, and six by the end of month 6
- *Objective 5:* To add one aircraft per month during year 2 for a total of 18 at year 2 end
- *Objective 6:* To form a marketing partnership with Airtran Airways

Each of these objectives has inherent business risks associated with it.

Objective 1. To obtain required DOT and FAA certifications on or before June 25, 2004.

Risk. That FAA certifications will not be obtained because of factors out of the control of the program manager, change in requirements, FAA delays, and the like. This is a good example of a business-wide risk that is addressed through a general contingency plan involving close and regular contact with FAA offices. But in the end this risk event could be pivotal.

Objective 2. To commence revenue service on or before June 25, 2004.

Risk. That revenue service cannot be commenced because of regulatory, equipment, or financial factors.

Objective 3. To raise sufficient seed and bridge capital in a timely fashion to financially enable these objectives.

Risk. That capital cannot be raised for the project; funding is critical. Contingency involves a wide sweep of potential financial backing.

Objective 4. To commence operations with two Canadair CRJ 700 regional jet aircraft in month 1, four by the end of month 4, and six by the end of month 6.

Risk. That operations will not commence at the level specified because of the lack of adequate aircraft support.

Objective 5. To add one aircraft per month during year 2 for a total of 18 at year 2 end.

Risk. That one aircraft per month is not possible because of equipment or financial factors.

Objective 6. To form a marketing partnership with Airtran Airways.

Risk. That the partnership cannot be consummated because Airtrain chooses not to enter into a binding relationship with one regional airline because of the restrictions the partnership would create on other marketing opportunities it is pursuing.

Program of projects

Good Flight Airlines is a new low-cost international consumer airline in its formative stages. It is being organized to take advantage of a specific gap in the low-cost international travel market. The gap exists in low-cost service out numerous markets in the United States.

The following program of projects outlines our initial objectives:

Program area 1

Start-up program. The start-up program of projects includes all program and project activities focused on commencing revenue service. The keys to success of the start-up program include obtaining the required government approvals, securing financing, hiring experienced management, and marketing—either dealing with channel problems and barriers to entry or solving problems with major advertising and promotion budgets. Targeted market share must be achieved even amidst expected competition. Product quality, safety, services delivered on time, costs controlled, and marketing budgets managed are to be reflected in the business plan (Tables 2.1, 2.2, 2.3).

- *Project 1:* To obtain required DOT and FAA certifications on or before May 11, 2004
- *Project 2:* To get airline certified (air carrier operating certificate)
- *Project 3:* To commence revenue service on or before June 25, 2004

TABLE 2.1 Net Present Value (NPV)—Program Area 1

Program Area 1—Start-up program
Candidate Project—Project 3: To commence revenue service on or before June 25, 2004.
(Cash Flows in Thousands of Dollars)

Year	Capital cost	Sales	Net income	Discount rate	PV
0					
1	13800*	110000			16%
2		216000			12%
3		240000			12%
4		250000			12%
5		260000			12%
1	(13800)	110000†	96200	.8621	82934
2		216000†	216000	.7972	172195
3		240000	240000	.7118	170832
4		250000	250000	.6355	158875
5		260000	260000	.5674	147524
				NPV:	732360

* Initial seed capital is to be attracted via a convertible debenture sold by *private placement.* This round of funding will have premium conversion privileges versus later rounds and bridge capital. The company has plans to proceed to a public offering prior to initiating revenue service. The expected proceeds from the private placement are expected to be $300,000 at seed stage, $3.5 million in bridge funding, and $10 million in I.P.O. proceeds (projected at $6 per share). Management cannot assure that I.P.O. will be available at the time desired and at the price sought.

† Sales figures are based upon load factors of 55 percent in year 1 and 62 percent in year 2. Second year revenues are expected to exceed $216 million dollars with additional aircraft and expanded routes.

TABLE 2.2 Weighted Scoring Model—Program Area 1

Program Area 1—Start-up program

Category	Weight	Project 1	Project 2	Project 3
Criteria support objectives	20	3	3	3
Good rate of return	20	3	2	3
Good net present value	20	2	2	3
Meets budgetary constraints	20	3	2	3
High probability of completing project	10	3	2	3
Good profitability potential	10	3	2	3

Project 1: 20(3) + 20(3) + 20(2) + 20(3) + 10(3) + 10(3) = 280
Project 2: 20(3) + 20(2) + 20(2) + 20(2) + 10(2) + 10(2) = 220
Project 3: 20(3) + 20(3) + 20(3) + 20(3) + 10(3) + 10(3) = 300
Based upon the weighted scores, Project 3 is the best.
1 = Unfavorable
2 = Satisfactory
3 = Favorable

Program area 2

Operational program. The *operational program* includes all program and project activities focused on starting commercial revenue service (Tables 2.4, 2.5, 2.6).

- *Project 1:* To raise sufficient seed and bridge capital in a timely fashion to financially operate our first revenue flight
- *Project 2:* To commence operations with two Canadair CRJ 700 regional jet aircraft in month 1, four by the end of month 4, and six by the end of month 6
- *Project 3:* To hire experienced management team, pilots, flight attendants, and baggage handlers; outsource aircraft maintenance, fueling, and food service
- *Project 4:* Purchase CMS reservations system; preferred over outdated Sabre or Apollo legacy system

Program area 3

Market analysis program. The *market analysis program* includes all program and project activities focused on evaluating performance on a month-to-month basis—initiating new services to future destinations, terminating unprofitable routes, and meeting our growth objectives (Tables 2.7, 2.8, 2.9).

- *Project 1:* Media executions will use local media, which is highly targeted and cost effective on a cost-per-impression basis.
- *Project 2:* Analyze traffic data to ensure profitability. Analyze local international traffic data to realize new markets for operations.
- *Project 3:* Air operations and reservations will be centralized and cost effective.

TABLE 2.3 Risk Matrix—Program Area 1

Task	Risk	Probability (%)	Impact	Severity	Contingency plan
Filing US DOT application	Application incomplete or missing required information	25	Delay projected commencement date	Showstopper	Resubmit application with DOT
Pass DOT fitness test part 1	Sufficient business and aviation experience	25	Delay projected commencement date	Showstopper	Ensure management team experienced
Pass DOT fitness test part 2	Review of operating and financial plans	25	Delay projected commencement date	Showstopper	Ensure seed and bridge funding are in progress or completed Backup funding programs
Pass DOT fitness test part 3	Applicants history of compliance record with DOT rules and regulations	10	Delay projected commencement date	Very high. Would have to realign management team	Should not pose a problem with intensive background check of management staff
FAA preapplication statement of intent	FAA preapplication not on file before US DOT reviews application	20	Delay certification	Medium	Ensure that this application is filed first
Incorrect data on DOT application	Markets served, frequency of flights, aircraft type inconsistent with first year revenues and expenses	30	Delay certification	Showstopper This will cause DOT to reject application	Ensure that application is in alignment with financials

- *Project 4:* Ongoing distribution strategy—establish own website with reservation, purchase, and payment capability.
- *Project 5:* Promotion strategy—employ public relations firm for both consumer and financial purposes.
- *Project 6:* Implement sales forecast and sales strategy.
- *Project 7:* Add one aircraft per month during year 2 for a total of 18 at year 2 end.

TABLE 2.4 Net Present Value (NPV)—Program Area 2

Program Area 2—Operational Program
Candidate Project—Project 2: To commence operations with two Canadair CRJ 700 regional jet aircraft in month 1, four by the end of month 4, and six by the end of month 6.
(Cash Flows in Thousands of Dollars)

Month	Capital cost	Sales	Net income	Discount rate	PV
1	330*	9200		12 %	
2	330	9200		12%	
3	495	9200		12%	
4	660	9200		12%	
5	825	9200		12%	
6	990	9200		12%	
1	(330)	9200†	9200†	.8929	7920
2	(330)	9200†	9200†	.7972	7071
3	(495)	9200	9200	.7118	6196
4	(660)	9200	9200	.6355	5427
5	(825)	9200	9200	.5674	4752
6	(990)	9200	9200	.5066	4159
				NPV:	35525

*Aircraft will be obtained on a *dry lease* basis (without fuel) from one of several aircraft lessors at an approximate cost of $165,000 per month. Generally, first and last month's lease payments are required in advance. Lease is usually a 5-year operating lease and most often qualifies as an expense item to the lessee.

† Sales figures are based upon 7 cents per *available seat mile* (ASM). We will achieve our target of 7 cents or less per available seat mile by a combination of cost saving measures:

Flight crew utilization of 60 percent above industry average.

Pilots and flight attendants will be deployed an average of 85 h per month versus an industry average of 50–60 h.

The company will realize additional savings in the insurance and benefits area by virtue of having fewer crew members.

Eliminating meal service in-flight will save approximately $3.00 per seat, per flight.

Operating one aircraft type will result in lower training costs for flight crew.

Analysis of business value

Selection of projects was also based upon the following mandatory tasks required to operate revenue flights:

Prepare a comprehensive economic and market specific feasibility study

Develop required operations and maintenance manuals and standards

Recruit, hire, and train executive and operations staff, for both permanent and temporary positions

Guide airline through *proving runs*

Provide all airline documentation (ticketing, IATA codes, computer reservation systems, schedule filing, accounting systems, and revenue management systems)

Choose facilities and equipment location

TABLE 2.5 Weighted Scoring Model—Program Area 2

Program area 2—Operational program

Category	Weight	Project 1	Project 2	Project 3	Project 4
Criteria supports objectives	20	3	3	2	3
Good rate of return	20	3	3	3	2
Good net present value	20	2	3	1	2
Meets budgetary constraints	20	3	3	3	2
High probability of completing project	10	3	3	2	2
Good probability potential	10	3	3	2	2

Project 1: 20(3) + 20(3) + 20(2) + 20(3) + 10(3) + 10(3) = 280
Project 2: 20(3) + 20(3) + 20(3) + 20(3) + 10(3) + 10(3) = 300
Project 3: 20(2) + 20(3) + 20(1) + 20(3) + 10(2) + 10(2) = 220
Project 4: 20(3) + 20(2) + 20(2) + 20(2) + 10(2) + 10(2) = 220
Based upon the weighted scores, Project 2 is the best.
1 = Unfavorable
2 = Satisfactory
3 = Favorable

TABLE 2.6 Risk Matrix—Program Area 2

Task	Risk	Probability (%)	Impact	Severity	Contingency plan
Unable to obtain aircraft financing	Seek other lenders	25	Delay projected commencement date	Showstopper	Seek other financing options
Aircraft availability 120 days lead time for additional planes	Adding additional aircraft at desired rate	25	Delay projected fleet expansion	Medium	Identify similar aircraft by manufacturer, which require no cross training
Aircraft registration	Delayed	15	Delay projected first revenue flight	Showstopper	Ensure aircraft registration filed with governing body ahead of schedule
Aircraft interior reconfiguration to company specs	Delayed	10	Delay projected commencement date	Showstopper	Should not pose a problem if aircraft are leased in timely fashion
Aircraft "proving run" on initial routes	Range, weight, and similar problems	20	Delay start of service	High	Proving runs should be completed ahead of start of service date
Aircraft painted in company livery	Delayed	30	Delay start of service date	High	Ensure that livery is applied prior to aircraft delivery

TABLE 2.7 Net Present Value (NPV)—Program Area 3

Program Area 3—Market Analysis Program
Candidate Project—Project 1: Media executions will utilize local media, which is highly targeted and cost effective on a cost-per-impression basis.
(Cash Flows in Thousands of Dollars)

Year	Capital cost	Sales	Net income	Discount rate	PV
1	180*	110000		12%	
2	240	216000		12%	
3	300	240000		12%	
4	360	250000		12%	
5	420	260000		12%	
1	(180)	110000†	109820	.8621	94676
2	(240)	216000†	215760	.7972	172004
3	(300)	240000	239700	.7118	170618
4	(360)	250000	249640	.6355	158646
5	(420)	260000	259580	.5674	147286
				NPV:	743230

*Marketing is targeted locally. The advantage of a local and highly identifiable market is that media selections can be limited in scope. There is no need for a national media program to launch Good Flight Airlines. The most effective media is expected to be outdoor billboards especially in Spanish. Other media will be local spot TV and highly visible programs such as local news and sports. Specifically Hispanic TV programming and newspapers will also be targeted. Advertising on MARTA trains and buses is also an option. Advertising also includes billboards, TV programming, and newspapers in Mexican cities the airline will service.

Contingency plans

Program area 1—Start-up program

Candidate project (project 3): To commence revenue service.

The risk matrix showed jeopardy and risk to the start-up program because of the application process to obtain the required U.S. Department of Transportation (DOT) and Federal Aviation Administration (FAA). The FAA preapplication statement of intent, FAA formal application, FAA inspection, and the DOT application to apply for air carrier certification need to be filed in some semblance of order. The FAA preapplication statement of intent needs to be on file with the local Flight District Standards Office prior to submitting the DOT Application for air carrier certification. Required filing fees must accompany all applications. Incomplete and erroneous data will result in applications being rejected from both agencies. This will lengthen the application process. The normal application process takes from 6 months to 2 years. The FAA application requires a fitness test of the air carrier including financial strength and executive management experience. Also, the FAA requires inspections during and after the application process including the proving flights. The proving flights ensure that the carrier can operate safely. The contingency plan for these risks are that

TABLE 2.8 Weighted Scoring Model—Program Area 3

Program area 3—Marketing program

Category	Weight	Project 1	Project 2	Project 3	Project 4	Project 5	Project 6	Project 7
Criteria supports objectives	20	3	3	3	2	2	3	3
Good rate of return	20	3	2	3	2	3	2	2
Good net present value	20	3	2	2	2	2	2	2
Meets budgetary constraints	20	3	2	3	2	2	2	1
High probability of completing project	10	3	2	3	2	2	2	1
Good profitability potential	10	3	2	3	2	2	3	3

Project 1: 20(3) + 20(3) + 20(3) + 20(3) + 10(3) + 10(3) = 300
Project 2: 20(3) + 20(2) + 20(2) + 20(2) + 10(2) + 10(2) = 220
Project 3: 20(3) + 20(2) + 20(3) + 20(3) + 10(3) + 10(3) = 280
Project 4: 20(2) + 20(2) + 20(2) + 20(2) + 10(2) + 10(2) = 200
Project 5: 20(2) + 20(3) + 20(2) + 20(2) + 10(2) + 10(2) = 220
Project 6: 20(3) + 20(2) + 20(2) + 20(2) + 10(2) + 10(2) = 220
Project 7: 20(3) + 20(2) + 20(2) + 20(1) + 10(1) + 10(3) = 200
Based upon the weighted scores, Project 1 is the best.

applications submitted to the governing agencies need to be checked and rechecked for completeness and accuracy.

Program area 2—Operational program

Candidate project (project 2): To commence operations with two Canadair CRJ 700 regional jet aircraft in month 1, four by the end of month 4, and six by the end of month 6.

Leasing or purchasing aircraft is the most difficult task in this program. Obtaining the required seed and bridge funding to finance the first two aircraft is vital to the success of this program. To dry lease aircraft (without fuel) most companies require the first and last month's payment. Also, to acquire new aircraft 120 days lead-time is required. Purchasing brand new aircraft will double or triple that timeframe. Aircraft registration, interior reconfiguration, livery application, and proving runs need to be completed prior to entry into revenue service. Various financing options need to be investigated along with a continuous search for investors. The tasks required after obtaining aircraft can be scheduled accordingly based upon delivery.

Program area 3—Market analysis program

Candidate project (project 1): Media executions will utilize local media, which is highly targeted and cost effective on a cost-per-impression basis.

TABLE 2.9 Risk Matrix—Program Area 3

Task	Risk	Probability (%)	Impact	Severity	Contingency plan
Identify sources in the market that might cause lower profits	Sources that will lower profits include competition, higher airport fees	25	Lower revenues that projected	Medium	Build to-be expected risks into financial plan
Focus on long-term profitability, not short-term windfalls	Government regulations can affect prices	25	Lower revenues if long-term not planned regardless of profitability	Medium	Marketing strategies that depend on price chasing have not been shown to be consistently profitable
Know what level of risk you are comfortable with	Inability to control market forces and difficulty in predicting those forces make marketing an inexact science	25	Revenues could fluctuate up and down initially until a more precise marketing plan is implemented	Medium	Marketing involves understanding your level of risk tolerance; it also involves a good understanding of your current financial position
Be willing to increase the number of skills in your marketing toolbox	Successful marketers are continually updating their abilities by learning new skills	10	Competition and "not knowing" could affect passenger bookings	Very high Could impact financial statements	Hire outside marketing firm
Develop an integrated management approach to your business	Marketing decisions should not be made independent of other airline business decisions	20	Financial plans can be affected due to the legal nature of marketing	Medium	Marketing decisions involve contractual agreements that have important legal consequences

The marketing program has the following risks which are sources that will lower profits—competition and higher airport fees. Government regulations can affect prices. Marketing strategies that depend on price chasing have not been shown to be consistently profitable. Inability to control market forces and difficulty in predicting them make marketing an inexact science. Marketing involves understanding your level of risk tolerance. It also involves a good understanding of your current financial position. Successful marketers are continually updating their abilities by learning new skills. Hiring outside marketing firms can help. Marketing decisions should not be made independent of other airline business decisions. Marketing decisions involve contractual agreements that have important legal consequences.

Other issues

Southwest Airlines is one of the most successful airlines in aviation history. When Southwest enters a market, it stimulates traffic by significant amounts, competes aggressively with the majors, and has never once claimed it has been the object of predation. It ignores travel agents and avoids flying into slot-controlled airports and other factors that the DOT and the General Accounting Office (GAO) consider vital to accomplishment. The current crop of start-ups has less than a 5 percent probability of becoming another Southwest. New entrant airlines don't need help failing—they do a pretty good job on their own.

- New entrants choose their routes and markets poorly. The average duration of the current crop of start-ups on a city pair where they retreat from the market is less than 6 months over the last 5-year period.
- New entrant airlines price below their costs. The RASM-CASM (revenue per available seat-mile subtracted by the cost per available seat-mile) shows that new entrants consistently price below their costs. At their current price levels there is little hope that their break-even load factors will improve or that they will become long-term viable airlines.
- New entrants often hire executives with demonstrated records of failure. Over 97 percent of the carriers filing for Chapter 10 bankruptcy during the 1990s had senior executives who had been involved in a previous Chapter 10. Over 75 percent had executives who were involved in two bankruptcy filings, and over 50 percent have had executives who have been involved in at least three bankruptcy filings. 15 percent of the carriers filing for Chapter 10 bankruptcy during the 1990s had senior executives who had been involved in at least four bankruptcy filings. One person has been involved in five airline bankruptcies.
- According to the airlines themselves, predatory practices by major airlines are not a factor in new entrant failure. The reasons most cited for bankruptcies are because of operational plans, excessive debt, escalating costs, inadequate traffic, and economic downturns.
- Southwest offers the advantage of high-frequency service to its customers. Southwest also has one of the highest average aircraft utilization ratios in the world. This ratio is a good measure of productivity. It shows how well the carrier uses scarce and expensive resources. As aircraft utilization goes up,

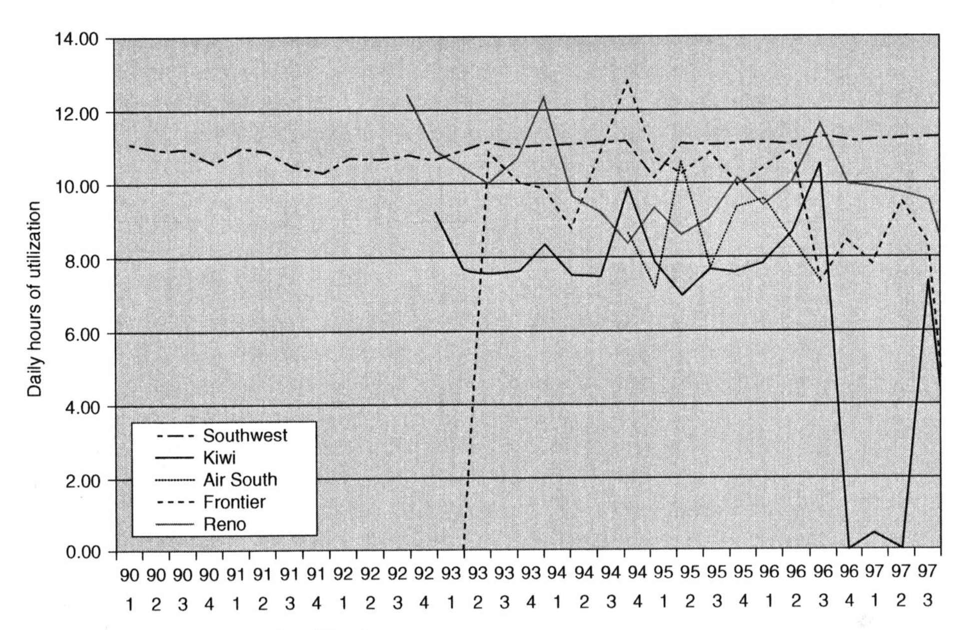

Figure 2.3 Average aircraft utilization.

unit costs go down. Marginal revenue also increases as a direct result of improvement in aircraft utilization. See Fig. 2.3.

This frequency allows Southwest to charge low fares consistent with low costs derived from productive assets. This is why Southwest makes money consistently. You would think this model would be imitated by the start-ups, but it is not. This is in strong contrast to the start-ups that come into many markets with low frequency. Low frequency causes unit marketing and station costs to be high as scarce resources are inefficiently utilized. Southwest has at least three times the frequency into markets that the start-ups have. It seems that high frequency is what consumers want.

- This is the most impressive fact about Southwest—they carefully manage and control their growth. One of the problems with the fast growth scenario is that management is unsure of costs and their pricing policies are not sustainable. The former chairman of People's Express blamed the downfall of his airline to the lack of revenue management.
- One more comparison shows that the start-ups are flying much longer routes than Southwest. The average stage length of current start-ups is about 33 percent longer, which is the reason why their yields are so low. This reflects once again on poor route selection. If new entrants were able to fly shorter segments at the same price level they now have, they would more likely get the revenue they need to succeed.

- The constant changes in operational plans by the current start-ups, which include changes in management with new entrant airline executives, whose average corporate life expectancy is less than 2 years. This last figure seems systematic of the problems that start-up airlines and businesses have in common. Poorly managed companies with constant turnover.

Comments on Risk Analysis

Demystifying risk in the portfolio process starts with the proposition that risk is one aspect of the strategic planning process and selection of programs and projects. As this case illustrates, the risk matrix process and separate risk scores are inputs to a decision to proceed that includes financial and alignment (weighted scoring model) issues that themselves involve risk. Such decisions cannot be reduced to quantified equations and mathematical relationships, or to so-called scientific calculations of risk probability and expected value. The point is that all front-end business planning is risk planning.

Chapter

3

Doable Tools: Applying Tools Strategically

There is a practical way to integrate risk into project planning and control and there are project tools that effectively support risk management. Once risks are identified as business risks, risk assessment and management are integrated into the project planning process. The purpose of this chapter is to illustrate the process. The process described below *embeds* risk into many of the traditional project planning and control steps, simplified and "demystified" for practical use.

Organizing a project from scratch and integrating risk involves seven basic steps:

1. Customer requirements
2. Work breakdown structure
3. Task list with estimated durations, linkages, and resources
4. Risk matrix
5. Network diagram
6. Time-based network diagram
7. Gantt chart (schedule)

We will use a building project as our case. The deliverable in this case is an office building and the customer is a real estate property manager.

Customer Requirements

Requirements are customer-driven and so are risks. A customer-driven project has the best likelihood of success simply because the process focuses continuously on defining and redefining customer needs, requirements, and expectations first and then defining the scope of work.

The customer requirements document is the project manager's definition of customer requirements, developed from information gleaned from the customer. The customer requirements document is not the same as the project deliverable document or scope of work. The requirements document addresses the customer's business setting and needs and expectations for a project solution.

The scope of work or deliverable document is the project firm's plan to address risk in the production of a solution to the customer need. This document is used to anchor the project tasks, but it is aligned with the customer requirements—and constantly realigned during the progress of the project.

In our case, the customer—a property manager—has a "vision" of success, which must be captured in the customer requirement document. This is not simply a building design and specification, but a description of the customer's vision of how the building will look and perform for its tenants, and produce profitability for its owners.

Work Breakdown Structure

First level: the deliverable

The first step in defining the work necessary to produce the deliverable is to complete a *work breakdown structure* (WBS) from the top (the deliverable) down to the third or fourth level of tasks. You do this in outline form. The top of the WBS is the first level of this *organization chart of the work.* It represents the final product or service outcome of the project, performing to specification and accepted by the sponsor, client, and/or user. In our case it is the building itself. The "building" at the top of the WBS implies a finished product accepted by the user or customer.

Second level: summary tasks

The second level across the organization chart of the deliverable includes the five or six basic "chunks" of high-level work that serve as the basic components of the project—the summary tasks which are integrated at the end of the project to complete the job. For our building project these chunks of work might include the architectural drawing, building supplies, ventilation systems, water, and electrical systems. (For a software project these chunks might include hardware platform, software, interfaces, training program, and financing. For a health management system they might include the clinic population, health information system, medical personnel, space, and equipment.)

It is important to see summary tasks not only in terms of producing the deliverable but also in terms of activities to address risk and contingency needs. This is where risk and contingency planning starts—in identifying and describing risks and what contingencies need to be scheduled.

Third level: subtasks

The third level includes a breakdown of the summary tasks outlined above into two or more subtasks, which would be necessary to complete to produce the second level summary task. For our building project, under the summary task "architectural drawing," this might include three tasks—get an architect, prepare preliminary blueprint, and check against standard blueprint template.

Fourth level: work package

The fourth level is another level of detail at the real tasking level—the work package. These are the individual tasks assigned to team members. For instance, breaking down the summary task "get an architect" to the fourth level, we identify nine work packages—build list of candidate architects, develop criteria for selection, screen candidates, interview candidates, conduct reference checks, compile candidate information, distribute candidate information, convene meeting, and conduct process of selection.

It is this last level that is used to create the task list and identify risks, the actual work assignments that will be necessary to schedule and the risks that the work will not make schedule, cost, and/or quality requirements. These are the "schedulable" tasks and these are the tasks that involve risk; each faces some constraint or resource problem that could create task failure.

The resultant WBS outline (can be shown as an organization chart) looks like this:

The building

1. Architectural drawings
   a. Get an architect
      (1) Build list of candidate architects
      (2) Develop criteria for selection
         (a) Risk action: involve the customer in developing criteria
      (3) Screen candidates
         (b) Risk action: conduct a peer review on candidates to offset biases in project team
      (4) Interview candidates, conduct reference checks
         (c) Risk action: confirm references with second opinion
      (5) Compile candidate information
         (d) Risk action: scrub information to assure credibility
      (6) Distribute candidate information
      (7) Convene meeting
      (8) Conduct process of selection
   b. Prepare preliminary blueprint
   c. Check standard blueprint template
2. Building supplies
3. Ventilation system
4. Water system
5. Electrical system

Note that risk actions are designed into the task structure for tasks which have been identified as high-risk in the development of a risk matrix.

Of course, in this case we have filled out only one summary task to the fourth level; all levels are filled out in a real project. The concept is that all the project tasks at the lowest level of the WBS "roll up" to produce the deliverable.

Task List

The task list includes the fourth level of the building project referenced. This step defines the "work" for several purposes:

- To serve as the basic definition of the "work" of each task, consistent with the definition of work in MS Project (Work = duration × resource)
- To serve as the basis for the network diagram—each task will be an arrow in the network diagram
- To serve as the basis for identifying risks—the first opportunity to identify high-risk tasks

Tasks are listed and durations for each estimated by the task manager who will be accountable for that task, along with dependencies (predecessors) and assigned resources (Table 3.1).

Next is a risk matrix, the fundamental risk template that captures the essential risk information in a project (Table 3.2).

Network Diagram

Having identified the basic tasks of this summary task, you now build a network diagram of this summary task, which is later integrated with other summary task diagrams to create the whole project network, as follows.

TABLE 3.1 Task List

ID	Task	Duration (total estimated elapsed time) (weeks)	Predecessor (linkage or dependencies)	Resources
A	Build candidate list	6	0	HR specialist plans department
B	Define criteria for selection	3	0	Project manager and architectural drawing task manager
C	Screen candidates	50	A	Architectural drawing task manager
D	Interview candidates	30	A, B	Project manager
E	Conduct reference checks	25	B	HR specialist
F	Compile information	35	C	HR specialist
G	Distribute information	3	F	HR specialist
H	Conduct selection process	3	E, G	Project manager

TABLE 3.2 Risk Matrix

Task	Risk definition	Impact (1 to 10)	Probability (1 to 10)	P × I (1 to 100)	Contingency plan
Build list of candidate architects	List does not contain high-quality, available architects	8	8	64	Focus on industry-leading architect and negotiate a contract
Develop criteria for selection	Criteria do not include key factors of success	8	2	16	Forget criteria for selection and go find the best architect in the field, as in above
Screen candidates	Screening process does not uncover weaknesses or availability issues	5	1	5	Forget screening and pursue best in class architect
Interview candidates, conduct reference checks		10	5	50	Don't interview any more candidates
Compile candidate information		10	1	10	Don't compile candidate information
Distribute candidate information		9	2	18	Don't distribute information again
Convene meeting		5	10	50	Have meeting, but focus on one target contractor
Conduct process of selection					Sole source

Start with a network template. Always start your network diagramming with a template or model of the "typical" network, and adjust it to the project you are planning. A typical template looks like this (Fig. 3.1), with three paths and parallel activities ending in one task.

Then tailor your model to your project as shown in Fig. 3.2.

Project paths

$$A, C, F, G, H = 6 + 50 + 35 + 3 + 3 = 97 \text{ weeks (critical path)}$$

$$A, D, G, H = 6 + 30 + 3 + 3 = 42 \text{ weeks}$$

$$B, E, H = 3 + 25 + 3 = 31 \text{ weeks}$$

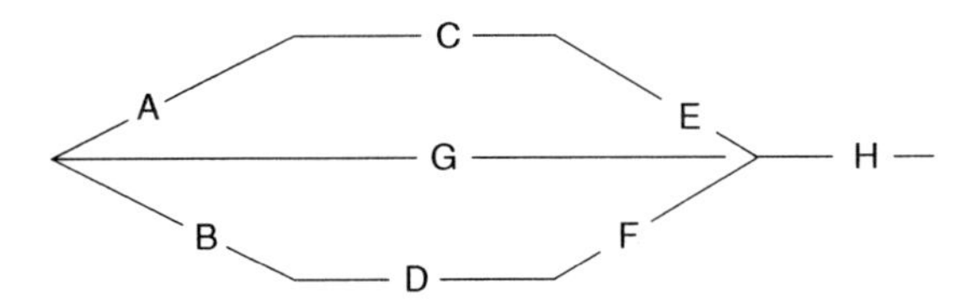

Figure 3.1 Generic network diagram.

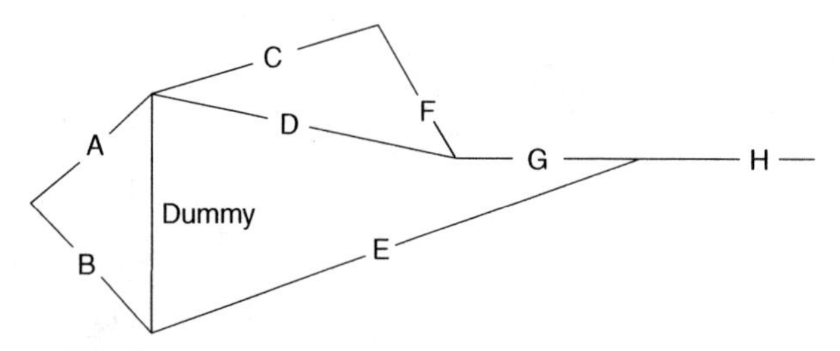

Figure 3.2 Tailored network diagram. (The "dummy" arrow connecting A and B is not a task but a link. This arrow shows that A and B are interdependent with D. D is dependent on *both* A and B, not just A).

Time-Based Network Diagram

Here you simply place the network diagram on a time-based graph. Draw the length of the arrows representing each task to equate with their actual durations as aligned with the bottom calendar of 97 days. Note that Fig. 3.3 shows *float*, or *slack*, the dotted lines that represent the flexibility in what *time slot* you determine to do the noncritical path tasks. Note also that the path, A, C, F, G, and H, a continuous arrow with no breaks, represents the critical path.

Analysis of early and late starts and slack

In order to determine what slack you have in the project plan to move tasks that are not on the critical path do an analysis of early and late starts and early and late finishes, and slack, as shown in Table 3.3.

Project paths

$$A, C, F, G, H = 6 + 50 + 35 + 3 + 3 = 97 \text{ weeks (critical path)}$$

$$A, D, G, H = 6 + 30 + 3 + 3 = 42 \text{ weeks}$$

$$B, E, H = 3 + 25 + 3 = 31 \text{ weeks}$$

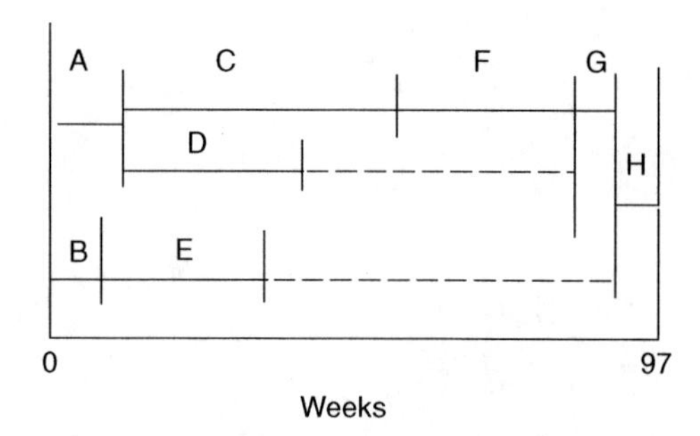

Figure 3.3 Time-based network diagram.

TABLE 3.3 Early and Late Start Analysis

ID	Task	Duration (total estimated elapsed time) (weeks)	Early start (week)	Late start	Early finish	Late finish	Slack
A	Build candidate list	6	0	0	6	6	0
B	Define criteria for selection	3	0	66	3	69	66
C	Screen candidates	50	6	6	56	56	0
D	Interview candidates	30	6	66	36	96	60
E	Conduct reference checks	25	3	69	28	94	66
F	Compile information	35	56	56	91	91	0
G	Distribute information	3	91	91	94	94	0
H	Conduct selection process	3	94	94	97	97	0

Gantt Chart

The final *Gantt chart* (MS Project) takes the task list information entries and builds a bar chart representing the whole project graphically, based on linkages and durations (Fig. 3.4).

This process, from initial requirement through the Gantt chart, is the core process of project risk management. This is where risks are captured and identified, as part of the project planning process.

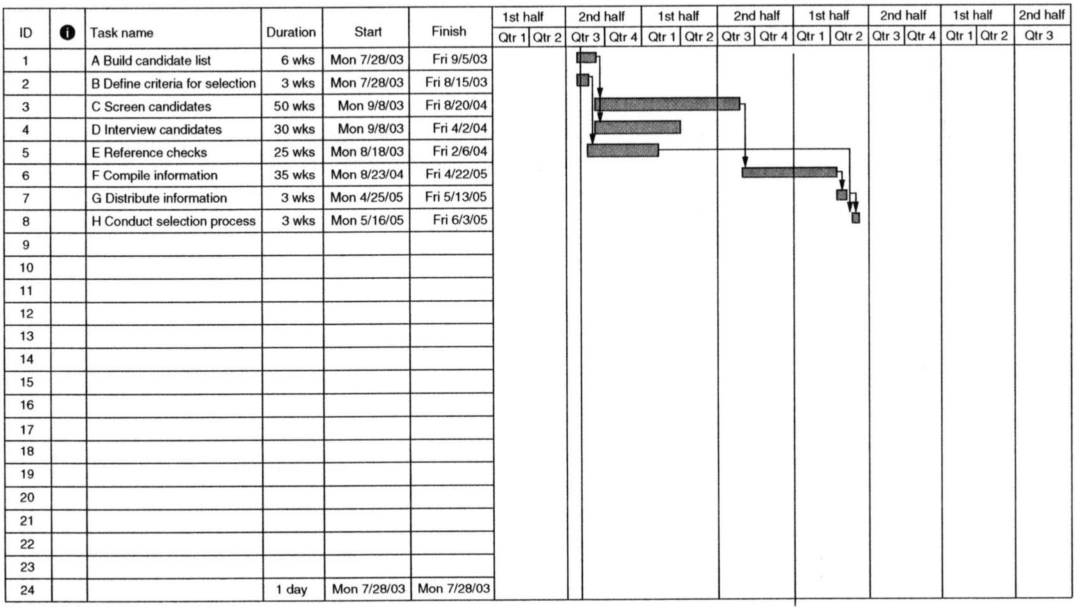

ID		Task name	Duration	Start	Finish
1		A Build candidate list	6 wks	Mon 7/28/03	Fri 9/5/03
2		B Define criteria for selection	3 wks	Mon 7/28/03	Fri 8/15/03
3		C Screen candidates	50 wks	Mon 9/8/03	Fri 8/20/04
4		D Interview candidates	30 wks	Mon 9/8/03	Fri 4/2/04
5		E Reference checks	25 wks	Mon 8/18/03	Fri 2/6/04
6		F Compile information	35 wks	Mon 8/23/04	Fri 4/22/05
7		G Distribute information	3 wks	Mon 4/25/05	Fri 5/13/05
8		H Conduct selection process	3 wks	Mon 5/16/05	Fri 6/3/05
9					
10					
11					
12					
13					
14					
15					
16					
17					
18					
19					
20					
21					
22					
23					
24			1 day	Mon 7/28/03	Mon 7/28/03

Figure 3.4 Gantt chart.

A Risk Story: Tradeoffs in Risk

Let's say that an electronics firm is facing a major decision in the use of an integrated printed wiring board (PWB) as part of its avionics instrument product line. The risks in that decision are inherent to the business itself and to any new product project. The background is instructive to the issue of demystifying risk and cost management.

The firm can gain a $500 price reduction from a new supplier if it chooses to purchase the new PWB that is assembled as a single unit, which appears to perform better than the old PWB, which is assembled from parts. But there are risks in the decision because at the same time that the new PWB can perform more effectively and meets government certification requirements, it has never been installed in a revenue aircraft and thus is still considered in the field to be in *test*. Thus there is some concern that customers will stay with the old PWB simply because of the current safety record of the old PWB in use.

Another complication in the project is the fact that the firm's manufacturing unit workforce is raising major concerns about the new PWB. The firm's production assemblers would not have a job should the firm go to the new PWB, simply because assembly and *bonding* of the card would no longer be necessary. The union representing 20 *bonders* has threatened legal action if any bonders are terminated because of this decision. Thus there is added risk and uncertainty in a major nonmonetary factor—workforce objection.

This is a good example of a project risk that is intertwined with a company-wide business risk—the probability of disruptive union action triggered by a major product development risk decision. If the project manager is accountable only for the delivery of the product prototype, then the risk of union action is of little interest in making the product decision. The project manager will proceed with the new PWB circuit board because it offers benefits in performance and cost reduction, and will take his or her chances with the marketing issue because he or she is not responsible for sales or cash flow. However, if the project manager is accountable for the whole cycle—product development, manufacturing, and successful marketing—then the PWB decision is liable to go the other way. The manager may not want to endanger the success of a proven product because of possible union and workforce disruption that could stop the production process.

Thus definition of the project (is it simply producing the prototype or is it getting the product manufactured and assembled and marketed?); its organization (who is responsible for what in that cycle and who do they report to?); and assignment of responsibility to the project manager (how is the project manager's performance going to be evaluated?); can all create the conditions for how tradeoffs of risk and cost are made.

Further, the business itself is vulnerable because of the precarious position of the manufacturing workforce. Without any alternative way to use the bonders, the company has no contingency for moving the bonders into other valuable roles and functions, thus a key product enhancement opportunity is lost.

The risk demystifying message is this: no amount of quantitative analysis and probability estimates will make this project management decision easier to make unless the organizational and accountability issues are resolved first. Even then the critical decision is subjective. The actual decisions are liable to be made in conversations among the project manager, the program manager, and stakeholders, and the "rank ordering" that results will develop a priority that will be clear to all.

Business and project risk

Thus project risk management does not start with the project; it starts with the business itself. As indicated before, it is quite apparent that many of the key forces in creating project risk are external, not internal, and are uncovered early in business strategic planning and environmental scanning. Many of the key factors in the success or failure of any project are the broad business factors for success and failure and the process of selecting the project in the first place.

The first step in project risk management is understanding the broad approach to the business you are in, the market, the strategy, the viability of the business organization and support system for project management, and the outcomes of SWOT analysis—strengths, weaknesses, opportunities, and threats. This is not a mysterious process that can be performed only by select corporate planners and sophisticated modeling. The process simply requires that the business purpose and strategy be made clear, and that *the risks and threats faced by the business are integrated into all project plans and systems.*

Market. Market analysis involves the investigation into potential markets and business opportunities. Projects interface directly with marketing in the sense that projects are designed to produce products and services that are consistent with the marketing plan. Marketing generates accelerated product development by spurring product concepts and making "deals" with customers and clients to deliver to their needs.

Projects implement business marketing and strategic objectives. Therefore, whether or not the marketing analysis is formal and documented, projects are typically conceived to implement the purpose and direction of the business. "Projects are what business does to improve." It is also possible that businesses take on projects to enter new fields and develop new business opportunities so the interrelationship between business strategy and project selection is actually reciprocal.

Client setting. The client setting is important because the boundaries of risk and uncertainty are set in the business setting itself. For instance, if a business is an exclusive provider of a particular product or component, the risks are less in any project involving that product or component than they would be if there were no exclusivity. A business that is meeting an increasing demand for a particular product but facing major competition will try to reduce costs and enhance price while maintaining quality and service. In that setting, a product development project that increases performance and cost is liable to be rejected as "risky" *in the context of that business setting.*

Strategic statement. Business makes strategic statements, either formally through written statements of strategy for investors, shareholders, and employees, or informally through the actions they take. It is important to recognize that most businesses operate with informal strategies, which are housed in the heads of its leaders, but nevertheless these strategies are clearly driving key business decisions. The truth is that most strategies are not written.

But a project manager dealing with planning decisions, risk, and cost must understand the business strategy—whether it is written down or not—and fit the project into that strategy because the risks the manager faces in the project are directly linked to the assumptions and conditions behind the strategy.

Strategic objectives. Strategic objectives are, in effect, statements of risk contingency—indications that the business sees key challenges in entering the marketplace and has developed an approach to addressing risk and uncertainty. Thus the whole strategic planning process and sets of statements of objectives is aimed at uncovering and addressing risk and uncertainty, in a sense, that downstream will provide the risk framework for projects designed to implement them. This way of looking at risk broadens the perspective of those who plan and direct project initiatives and also provides a useful way to "alert" project teams as to the risks and uncertainties that they face.

SWOT analysis: risk identification

Project risk identification, assessment, and management starts not with a project, but with business planning. Risks are typically identified in the broad business strategic planning and thinking that goes on to direct a company toward its potential markets and customers and toward appropriate products and services. Demystified, risk planning is a business function that identifies barriers and challenges for the company as it enters a market. The results of broad business planning provide the wherewithal for individual project risk assessment, which narrows down business risk into project risk. This translation is only possible if the business actually has a planning process—not necessarily a formal documented process but a way of thinking about the future of the company and its markets.

Strengths. The company looks at its competencies and its core capabilities and identifies its current performance advantage, its differentiators. These strengths are an important signal to the project manager of its history in overcoming risks and uncertainties in earlier improvement and product development projects.

Weaknesses. The company looks at its weaknesses in terms of its inability to overcome risk and uncertainty in past initiatives and identifies internal improvements that it must make to address them.

Opportunities. Opportunity is the converse of risk; there is no risk if there is not opportunity on the other side. The reason a company is undertaking a project is to "jump out" in the market to generate demand for its products and services and to face the possibility that it will fail because of competition or because demand was not there in the first place. Thus a project is a risk contingency, which, if overcome, becomes an opportunity.

Threats. Threats are risks, simply put. In other words, threats, such as competition, technological change, economic crisis, and financial difficulty, all constitute the risk factors that the business faces and thus the risk factors that individual projects face as well.

Weighted scoring model. We use the weighted scoring model to select projects based on their relative scores against strategic objectives, but what we are really doing in this process is evaluating contingency plans to address risk inherent in various candidate project plans, costs estimates, and cash flow forecasts.

Customer and client risks

Client risk management issues. It stands to reason that the client faces risks and the business and project objective is to remove or reduce the risk faced by the client. For instance, if I am developing an electronic instrument that provides a safety margin in an automobile, I am directly serving the client's interest in reducing risk associated with the automobile. While that is a simple and not very profound finding, it is surprising how easy it is to lose sight of the fact *the project risk management is inherently the process of managing customer risk.*

Project deliverable impacts. A differentiator for any project team is the capacity to anticipate impacts of project deliverables in terms of risk and cost and to offset them. That is what projects do fundamentally.

Partnering in risk management. Partnering in risk management aims at spreading risk and uncertainty out among the stakeholders so that no one will bear an inordinate cost should the risk impact success. Thus partnering proves that risk is shared between project company and client, between supplier and customer.

Project risk management: processes

Selecting the right projects. Projects are selected for a company portfolio of projects based on several factors including risk and cost. The review of risk at the selection process involves "order of magnitude" thinking about a project, for example, at the broadest level what are the factors involved in project success and failure and what are the probabilities of their occurring in this project? How can they be offset?

The analysis of risk in project selection involves three basic issues:

Financial risk. What is the probability (0, 25, 50, 75, or 100 percent) that the project estimates for financial performance are accurate? What is the risk of cost variance at the end of the project?

Schedule risk. What is the probability that the task duration estimates are incorrect and that the project will experience negative schedule variance?

Quality risk. What is the probability that the quality of the product or service, either in terms of the product specification itself or customer satisfaction, will not be delivered?

Managing by projects

Project life cycle. Risk is inherently tied to the timing and life cycle of the project because risk probabilities and impacts change over that life cycle. A minor, low-ranked risk at the beginning of the initiation stage can become a highly ranked and severe impact risk later in the project life cycle if not attended to. On the other hand, risks too early attended to can create needless effort and cost before their real implications for project success are fully known. Thus the effective project manager finds the appropriate "window" for risk response. That window is the decision point, or range, in the project where the risk contingency must be implemented to avoid schedule and cost impacts.

Initiation. The initiation stage is where the overall project risk is conceived, dimensioned, and described to develop an "order of magnitude" grasp of project risk. The kind of thinking and conversation that ensues in this phase addresses the *potential* of a new product or initiative, the *prospect* of customer and market demand, and the *competitive and risk* issues inherent in an endeavor still in its initial conceptual stage. At this point, the project manager and the planner are dealing with broad, "macro" issues such as described below for a variety of projects:

1. For a software project, whether the software deliverable would be new and untested or simply an enhancement or recurring production of software already in place on a similar platform. This project-level risk assessment improves on any business-wide strategic, marketing, technology, and environmental scanning information already available. In addition, the company's capacity to deliver the product to meet customer requirements is also reviewed to ensure that the risk of overcommitting the company and its resources is avoided. This where the tradeoffs are made between risk and opportunity, between cost and payoff, inherent in a new project, where its conceivers *play with and qualify* their misgivings and excitement about a new

effort. It is here that major business risks are first identified and linked directly to the likely outcomes of a particular project or product development process. Conditions and factors of success are rolled over to their worst case counterparts and potential customer requirements are reviewed to focus in on whether a customer would know what he or she really wants and needs—and act on it in the marketplace.

2. For a construction project, whether the building product is within the capacity and resources of the company, whether the customer knows the requirements, and what kinds of seasonal and other challenges lay ahead. Major risks in any construction project lie in the economic and social setting of the proposed investment and require analysis of a series of indicators around building vacancy rates, prices, and forecasted trends in demand.
3. For a telecommunications project, the broad issues of customer requirements, infrastructure and support, clarity of specifications, material costs, and other related cost issues are reviewed. The telecommunications risks and threats lie in the technology itself and the prospect of life cycle performance before being overtaken by another more effective technology. Cash flows are derived from best and worst case scenarios involving alternative telecommunications system life cycles.
4. For a health services project, the initiation phase brings out health technology and service developments, clarity of customer requirements, nature of the health services clientele, and probabilities of government assistance and regulatory activity. Early views of health services investments are fraught with the risks of technological and regulatory change, thus initiation requires a heavy dose of research and subjective judgment about change. Risk is a function of change since it is dimensioned uncertainty about whether a given effort will change a system or its performance as predicted and planned. The risk is that the intended change will not occur as predicted and further that whatever change occurs as a result of the project will create negative consequences for the client and the client's system, whatever it is.
5. Finally, the concept of order of magnitude costing applies here because it is in this phase that a prospective project cash flow is estimated along with costs. It is here that the initial level of profitability is estimated at a gross level, using whatever parametric indicators are available for the given industry. For instance, for an avionics instrument, the prospect of successfully marketing a new digitized instrument for a business aircraft is based on an industry working standard that it takes about 18 months to develop a new instrument and 5 years to recover costs.

Planning. Planning includes the development of project planning documents, scope of work, customer requirements document, schedule, budget, risk management plan, product development process, and project monitoring plan. The significance of the planning process for risk management is that risk is best addressed in the planning process simply because it allows time for assessment and contingency plans to be integrated into the baseline project schedule and budget.

Execution. Execution is the implementation phase once a baseline schedule is completed. Execution begins when work is authorized to begin and the project schedule is underway.

Control. Control is the project-monitoring phase during which performance indicators are monitored and the project is controlled by corrective action. The significance of control for the risk management process is that risk is built into earned value determinations. Earned value indicates variances from schedule and cost estimates, indications that risk may be at work in a project. Investigation of root causes involves first focusing on risks and risk events that were anticipated in the project planning process.

Closeout. Closeout is the termination phase in which a project is closed along with financial books.

Scope. The scope of work describes the work to be performed to produce the deliverable. Scope addresses customer requirements and deliverables, approach to producing the deliverables, and estimated schedule and cost estimates.

Time. Scheduling determines the project life cycle and delivery date for the deliverables. Time is an outcome of the planning process, not an input (despite the tendency of customers to dictate delivery dates without understanding the project process).

Cost. Cost is capture bottom up and top down. Bottom up estimates fixed and labor costs from individual tasks at the work package or level of effort, and rolled up. Top down applies parametric indicators, such as plan your building around $40/ft^2 as the industry standard for this kind of building.

Risk planning. Project planning and risk planning are related in the sense that project planning documents and deliverables incorporate risk and risk contingencies. But risk planning has taken on another slightly different meaning in the new PMI *PMBOK* document. Risk planning is seen by PMI as *preparing the organization and its support systems for risk management.* This emphasis gives special attention to the need to build the cultural underpinnings and support systems for risk management before top management can expect its project teams to address risk.

Risk planning involves the following outcomes:

1. *Development of a risk management policy and procedure system.* This involves the "institutionalization" of risk management into the policies, manuals, and procedures of the company. While stated policy does not always influence actual work setting behavior, it is important for the company to go public with risk as a priority management ethic.

2. *"Walk the talk" programs—orientation of top management to the integration of risk into day-to-day work and communications.* This involves assuring that top managers actually address risk in their project review meetings, customer communications, and performance reviews.
3. *Training and development programs for risk.* The company must have a risk management training program backed up by manuals and web-based training in risk assessment and response.
4. *Network and web-based information system development for project risk information management.* Since a company will want to address risk consistently throughout its project portfolio, a network should be earmarked for support to project managers, providing risk templates, forms, and data formats.
5. *Project management office support.* A separate staff serves project managers with standard and best practice, project review formats, monitoring data and analysis, and risk management services.
6. *Standard work breakdown structures.* The more the company can move toward a standard work breakdown structure, the easier it is to inculcate risk management practice into the scheduling and execution processes. Standard WBS formats will provide for risk contingencies and risk-based scheduling using MS Project PERT analysis or other software tools.

Quality. Quality is addressed in the project planning system first by assuring a firm grasp of customer requirements. This involves making sure the customer's needs and expectations are documented in the requirements document since in the end the customer actually determines the quality of the project outcome.

Second, quality is addressed in the application of quality assurance and quality control processes, ISO standards, and process improvement initiatives. Quality can be defined as the response to risk in the sense that quality issues stem from high-risk projects and project tasks. Therefore, it is important to address quality through the risk management process to produce a project system that protects against quality defects, variances, and high appraisal and inspection costs.

Human resources. The human resources element of project management involves the development of proficient workers and high-performance teams, focusing on strong support of the human resources staff. This means that employees who are comfortable with their work settings and employee support systems are more likely to seriously address good planning practice and protect the company from risk. As employees become more disengaged from the human resources function, they are liable to be increasingly alienated from the company and therefore be less interested in protecting it.

Communications. Risk is communicated in the project planning process through regular exchanges on risk in project reviews and task assignments. But the essence of risk communication is the ability to keep all stakeholders informed on risks so that they are not surprised by shifts to contingency plans, which push out schedules and increase costs and budgets.

Procurement. In any contracting system the contract is designed and managed with the purpose of minimizing risk and maximizing risk-sharing arrangements. This means that fixed price contracts are favored over cost reimbursable contracts in most cases because fixed prices shift the risk to the contractor.

Integration. Integration involves the mixing and matching of project components and parts to create the whole. Technically, integration is a systems function, but in terms of risk management the integration function means that risk is *embedded* in the process. For instance, in producing a product the company incorporates technical checks and balances in the system to offset the likelihood that product quality has been compromised.

Triple constraints

In a way, the so-called *triple constraints* of quality, schedule, and cost are not really constraints, they are outcomes of constraints. Constraints are resources, technology, risk, and project processes. People make it possible to produce project deliverables by managing these constraints to optimize schedule and costs while meeting quality and performance requirements.

Project Manager's Roles and Responsibilities in Risk Management

Leadership. The leadership function in risk management includes all the core leadership skills, e.g., vision, team development, giving purpose and direction, along with a strong sensitivity to risk. Leaders ask questions, but do not necessarily provide answers. The role of the leader in a risk management process is to ask the right questions, pose the right issues, and inspire the project team to come up with solutions and opportunities. Risk provides the leader with a conceptual framework for posing project issues in terms that can be incorporated into risk planning, matrices, and assessments.

Motivation. The classic role of motivating is actually a process of integrating individual leadership with a project work setting in which motivation is self generated. In other words organizational leadership cannot be very effective in motivating a project team unless the leader has designed the work itself to be challenging to the team. It is the work that motivates teams, not leaders alone. So it is the work that needs to be designed to be challenging. And since opportunity and challenge come from *risking yourself together to overcome a risk,* project teams typically are motivated by risk that is manageable.

Keeping risk manageable. Leaders keep risk manageable by framing a project that is feasible but that stretches the project team enough to represent a meaningful barrier. Overcoming that barrier then becomes the source of recognition and achievement and more motivation to go on to other challenges.

Facilitator/Manager. The facilitating gift is the capacity to guide a team to high performance by orchestrating the dialogue without dominating it. This means that facilitators must stay engaged with the team and keep the team moving by raising issues and challenges, but keep disengaged from the solutions and outcomes as much as possible. Solutions and outcomes are to be addressed by the customer in the context of customer requirements; the facilitator is interested in spurring the team to a high level of achievement, but not necessarily interested in guiding the resultant solution.

Work Breakdown Structure, Again!

Description, purpose, and use in decomposing a project. The reason a WBS is a necessary part of risk management is that the WBS defines the risks in a project at a level that the risk can be identified, described, and assessed. Thus you can look at the WBS as the structural support for risk management in the sense that it "rolls out" the work so that everyone can see the work in small enough chunks to ascertain the risks inherent in completing it. If a WBS misses a major chunk of work, it may also miss a major chunk of risk, so it is important for the WBS to define all the work.

Activity definition. Activities are project tasks that create outcomes, consume resources, and must be scheduled. Activities are differentiated from milestones, which are points in time, which mark significant endpoints or intermediate products of activities. Defining the activity—not just identifying it in a one-liner—thus is a critical step in risk management. The more detailed the activity the more clear the risks inherent in the work will be.

Resource planning. Resources are constraints and thus are the generators of risk. According to the theory of constraints, resources are the major sources of project failure or success. Using the "article chain" concept, the project manager is advised to trim estimates down to "bare bones" so that *buffer* time can be doled out when necessary, and to concentrate only on the key constraints in the project. Inherent in the theory is a major criticism of micromanaging all tasks as if they were all of the same importance. Isolating the few high-risk tasks gives the project manager focus.

Cost estimating. Cost estimating is a bottom-up and top-down process, but it begins with the WBS and builds labor and fixed costs as part of the scheduling process. At the heart of the cost estimating process is the definition of the work, with work equaling resources multiplied by time. In other words, a task might be defined as a "five person week job," that is, given the nature of the defined work it is estimated that it will take five people 1 week to do the task—a 200 h job—*if they are all working full time.* If less than five people are working on the task, or some are not full time, it will take longer, thus the duration and resource levels are traded off with each other. The work itself and its definition stay constant unless changed by a new estimate of the work. Costing out a job that has been defined to this detail is relatively easy since it includes the cost of full time staff working for 40 h.

How does risk figure into cost estimating? Risk is a driver of cost since its impact is to extend the work beyond the expected duration to a more pessimistic duration based on the potential occurrence of a risk and the time and cost of implementing a contingency action. Thus cost and risk are associated with each other; various levels of risk create different costs.

Budgeting. Budgeting is the allocation of available resources to projects based on priorities. While cost estimates are useful inputs to budgeting, the company decides where to put its money based on its *portfolio* analysis, using cost estimates.

Progress reporting. Progress reporting is accomplished through earned value, e.g., schedule and cost variance, as well as by risk assessment.

Coding of work breakdown structure (WBS) elements. The estimation of cost from the WBS includes the cost of all tasks and risk contingencies built up from the bottom of the WBS.

Relationship to cost accounting. Risk impacts on project cost are captured in the budget and since the WBS is coded to the task level, the cost of all risk contingencies should be documented in the cost accounting system.

Activity-based costing. The purpose of activity-based costing is to measure costs and therefore profitability based on the cost of time. This leads to accuracy in cost tracking as well as measuring resource capacity excesses and constraints. It also helps decide what costs contribute to profitability. The costs of risk mitigation are attributed to appropriate project tasks and subtract from project margins.

Relationship to responsibility matrix and organizational structure. The responsibility for risk is shared in a matrix between functional and project managers. The responsibility for process and technical risk is with the functional department and that for deliverable schedule and cost risk is with the project manager. The organizational structure should be clear on roles and responsibilities so that risk accountability can be assured.

Estimating

Estimating is the process of estimating schedules and costs based on a proposed deliverable specification. There are different types of estimates:

Order of magnitude/conceptual. Order of magnitude is "ballpark" estimating, getting a handle on the general cost based on all task work and expected risk mitigation costs. This estimate is general and helps to *scale* the project initially as complex and big, mid-sized, or small—in the context of the company's capacity and past work.

Budget/parametric. Parametric costing depends on the availability of parameters for scaling costs, e.g., the cost/square foot of building a health clinic building.

Definitive/detailed. Definitive is a detailed costing based on individual work packages and levels of effort at the individual and team level. This is the final budget against which cost variance is measured, including overhead and general and administrative costs.

Project type/industry. The estimating process differs depending on the industry. Estimating a construction project is typically at the definitive and detailed level while estimating an information technology project is often at the budget and parametric level because of uncertainties and risks.

Time

Activity duration. Activity duration estimates are based on information from the assigned team member, parametric data, and can be in three dimensions—expected, optimistic, and pessimistic—for PERT analysis.

Expert judgment. Expert judgment is used in providing order of magnitude and parametric budget estimates.

Analogy. If there is an analogous project to the target project, then analogous cost data are used, as in parametric costing.

Productivity rate. Productivity rates are useful in estimating the impacts of risks on costs and schedule because of the rate at which deliverable components and pieces of work can be produced. For instance, if a given report typically takes 1 week to design, prepare, and publish, then two reports can be done in 2 weeks or less.

Contingency time. Contingency plans take time and financing, so estimating the cost and schedule impact of contingencies should be part of the regular project planning and scheduling process.

Risk-based scheduling

Calculating a risk-based schedule: PERT analysis. We deal with MS Project in more detail in Chapter 7, but here is a summary of risk-based scheduling.

MS Project has the capability of calculating a risk-based schedule and task durations using the PERT tool. This is based on estimated expected, optimistic, and pessimistic task durations.

The purpose is to use risk assessment and analysis information to calculate a *risk-based* project schedule in Microsoft Project. The risk-based schedule is calculated from your original project schedule, but uses your weighted estimates of three possible task durations (expected, pessimistic, and optimistic) to come up with a new project schedule. The new schedule is calculated for individual tasks and "rolled up" to the whole project.

The risk-based schedule is usually a better schedule estimate than your original one because it reflects your best estimates of what could go wrong (risk) and what could go right (controlling risk).

Procedure

1. Prepare your regular project schedule using Microsoft Project using your best estimates of task durations and linkages. This project schedule does not reflect any risk assessment.
2. Prepare a risk matrix using the work breakdown structure (WBS) and the project schedule and rank *all* project tasks in terms of risk, designating them "high, medium, or low."
   a. A high-risk ranking shows a high probability (>50 percent) of the risk actually occurring, *and* that the risk will have a relatively severe impact on schedule, cost, and/or quality.
   b. A medium-risk ranking implies less probability (<50 percent) of happening and less schedule impact
   c. A low-risk ranking implies very low probability (<10 percent) that the task will occur and low impact.
3. Select the five highest-risk tasks (or more if you have more tasks that present risks that you want to reflect in your risk-based schedule).
4. Calculate the risk-based schedule: Your objective now is to calculate a risk-based schedule by taking each of the five highest-risk tasks and calculating a risk-based duration for each. Using Microsoft Project, here are the steps:
   a. Pull the PERT Analysis Toolbar up from "View."
   b. Highlight one of the high-risk tasks on the Gantt chart.
   c. Go to the PERT Entry Form and enter your duration estimates for that task for three scenarios—expected (use the duration in your original schedule), pessimistic (worst case impact if risk occurs), and optimistic (best case, all risk controlled with no impacts).
   d. Then use the PERT Weight button to set the weights for each scenario (weights reflect the probability that a given risk and impact will happen). Microsoft Project uses a total weight scale of six points; your job is to divide the six points up among three scenarios—expected, pessimistic, and optimistic. Note that the Microsoft Project "default" is 4 for expected (based on the high probability that the actual duration will fall somewhere between the two extremes) and 1 each for pessimistic and optimistic. But you may want to change those weights based on your estimate of the relative probability that a given scenario is going to happen.
   e. Once you have entered weights, go to the PERT Calculation button and calculate the risk-based duration for that task, based on your inputs.

f. Now click the PERT Entry Sheet and you will see the newly calculated, risk-based duration for the task compared to the three scenario durations (expected, pessimistic, and optimistic).
g. Now repeat this procedure for the remaining high-risk tasks.
h. The resulting "rolled up" schedule is now a risk-based schedule, reflecting a new project duration.

Dependencies

Risk and activity sequencing. It is important that the project schedule reflect the correct linkages. There is considerable risk inherent in these linkages since in a complex project these linkages change. Dependencies identified at the beginning of the project sometimes disappear as team members consult and collaborate before the tasks begin. And, conversely, some tasks not initially linked develop dependencies because of unanticipated developments in the project.

For instance, one design and one integration task might be initially linked because of the obvious need to design before a system is integrated. But as the team collaborates, it is probable that integration starts before design is completed and may go on in parallel. The risk is that because of Murphy's law (work expands according to the amount of time available to do it) the original task durations may be too long but are never changed. On the other hand, two project tasks, such as design and reporting progress, are not originally linked but later become interdependent because of an unanticipated request for a special report to a stakeholder on design review results.

Cost

Direct and indirect costs. Direct costs are costs of labor and other costs applied to the project by those who add value to the deliverable. Indirect costs are support costs—overhead and general and administrative costs. The risk inherent in costing is that the overhead and G&A costs are underestimated and the customer is surprised with invoices that exceed the expected costs.

Fixed and variable costs. Simply put, fixed costs are costs of capital, equipment, and space, and are entered as fixed costs into MS Project in building the project cost estimate. Variable costs are direct costs of labor, those that vary over time. The risk in fixed cost is the probability of misestimating fixed costs and the possibility that while fixed costs can be prorated over the life cycle of the project, cash flow is drained when the costs are actually incurred.

Basis for estimates

Internal accounting records. There are inherent risks in using internal accounting records as the basis for estimates. Each project is different and unless the project was exactly like the one under review, there is a high probability that past records will not be a good guidepost. On the other hand, internal accounting records are useful on the actual cost of labor and capital equipment.

Project team knowledge. There is risk in assuming project team knowledge about costs that may not be relevant to the new project. The best approach is to seek team insights, but to temper them with the perspective of a project manager.

Estimating manuals and databases. For deliverable components, estimating manuals can be useful as parametric guides, but again the risk is that manuals will not have accurate data.

Construction. In the construction industry, there are many sources of risk and cost data, from such organizations as The National Association of Home Builders. Parametric data are available on construction unit costs ($/sq ft) and supply and contractor risks and contract types.

Software development. Software development creates a unique set of risks and costs associated with this business. Since software development involves creative work in design and coding and faces major challenges in integration and platform options, the business does not have reliable cost and risk data and experiences major project delays.

The Software Engineering Institute is the major source of practical methodologies and risk and cost data. SEI defines a successful risk management practice as "one in which risks are continuously identified and analyzed for relative importance. Risks are mitigated, tracked, and controlled to effectively use program resources. Problems are prevented before they occur and personnel consciously focus on what could affect product quality and schedules."

The lack of focus on risk in software development has created major delays and cost overruns because of the lack of top management support, failure to gain user commitment, misunderstanding requirements, lack of adequate user involvement, failure to manage end-user expectations, changing scope, lack of technical skills, new technology, and staffing inadequacies and conflict.

Typical risk and cost issues are grounded in technology issues, e.g., two computers having different architectures that interpret a designated protocol differently.

Learning curves. As an organization gains competence in a given project arena, project life cycles are shortened. Performance time and experience can be represented by a learning curve relating the direct labor required to experience gained in past work.

Parametric estimating. Parametric estimating, that is estimating based on unit measures, such as $/sq ft, creates risk in the sense that these are median measures from a wide variety of similar projects and are not reliable for unique projects.

Issues

Poorly defined project scope. Since there is no industry standard on what is involved in a scope of work, the quality of the scope is typically measured against the number of changes or scope creep the project experiences. Risks and cost/schedule impacts are widespread in most project areas, particularly in software development and system developments. *PMBOK* states that the scope of work should include all the work necessary to complete the product to meet specifications, e.g., the WBS is a core element in the *PMBOK* framework. To the extent that the WBS covers all the work and by reference all the components of the product, and risks and costs are estimated from that WBS, the scope of work is the major anchor and stability factor in project planning and control. The potential for scope creep is a major generic risk in any project, and the best contingency is to assure that a WBS is prepared that is thorough from the perspective of the project team, stakeholders, top management, and the customer.

Major omissions in tasks/activities. If tasks are left out of the WBS and scope, then adding them requires a change order and rescheduling and rebudgeting of the project. Project managers cannot accept major additions or changes based on new work unless the project team is clearly at fault in missing an important piece of work.

Optimistic time/cost estimates. The natural tendency is to be optimistic in estimating schedule durations and costs, thus the risk is that the schedule is not risk-based. To offset that tendency, the project manager should use the outputs of the risk analysis process and risk matrix content to estimate a "pessimistic" duration option in the MS Project PERT analysis and then to calculate a risk-based schedule on that basis. The difference between the expected and pessimistic durations is a "buffer" in the sense that the term is used in the theory of constraints that should be controlled by the project manager.

Project Financial Perspectives

There is risk in the misuse or inaccuracy of financial analysis tools, and more importantly of the inputs or assumptions used in the application of those tools.

Concepts associated with interest rate; time value of money; simple interest; compound interest; discount rate; minimum acceptable rate of return (MARR); present, annual, and future worth (PW); and various rate of return and tax methods should be peer reviewed by an internal or external accounting consultant before being applied to projects. The risk is that rules of thumb and industry practice may be ignored.

Evaluation tools and techniques. Projects can be evaluated by several tools, including break-even analysis, payback period, replacement analysis, economic life, and feasibility studies. But the real issue in project evaluation is the longer term viability and value of the deliverable in meeting business and financial objectives, and the risk is that short-term, internal corrective actions during project delivery will be short sighted and focused on narrow schedule and cost issues, and not on long-term success.

Project budgeting

Budget inputs. Budget inputs include the risk management plan as well as the traditional project tools such as WBS and schedule. The risk in budgeting is the same as the risk experienced in scheduling—the tendency to be too optimistic. One way to address this issue is to perform a PERT analysis on a task using dollars instead of time units and to calculate a risk-based budget using that tool.

Preparation approaches

Top-down (strategic, tactical). Risk and cost are approached in a similar way, first from the top-down approach to develop "chunks" of risk and cost associated with various large packages of work, then to proceed to bottom up to refine the estimates.

Bottom-up (WBS-based). "Bottom-up" means taking the lowest level of the WBS (at least four levels down), estimating cost and risk, then rolling the budget up and preparing a risk matrix to rank risks with direct impacts on cost.

Iterative (combination of both). The iterative approach simply assumes that both approaches are helpful in developing a project budget and risk plan.

Estimates plus contingency

Refine estimates. Estimates are refined at the definitive level (from the budgetary level) by relating costs to specific measures and task durations, and to arrive at an estimate that can be used to propose a budget to a sponsor or investor.

Risk considerations. Risk is integrated into the refining of estimates by calculating a risk-based budget and schedule and comparing it to the original estimates, and making sure that all contingency plans have been estimated and integrated into the baseline schedule.

Issues

How much contingency. Contingency can be seen in the framework of critical chain theory so that the costs of the pessimistic option, both time and money, are earmarked as contingency funds and tapped by the project manager for allocation according to need.

Who controls contingency. The project manager controls contingency to avoid a multitude of contingencies being built into every task estimate, thus suboptimizing the project.

Management reserve

Project financing. There is substantial risk in the financing of a project, and more importantly in the financing of the parent company. Issues such as internal resources, commercial credit, equity issues, venture capital, government grants/subsidies, and capital rationing govern the viability of the company itself and its capacity to finance a project up front. To share risk, many companies are going to earmarked venture funds for new product development so that sponsors' funds are bounded by a project to avoid cross subsidy.

Insurable risk. Insurable risk is risk that is determined to be insurable on the open market and which is not self funded. Insurance is a vehicle to share risk and to outsource risk that has a high probability and impact severity.

Identification of project risks. Project risks will surface in a number of ways.

Internal. Internal risk is risk that is organizational and system-related, and that poses challenges to the company itself to support successful project management.

Technical. Technical and technology risk can be managed using reliability and testing methods, which must be built into the project itself. Technical risk is handled by embedding testing in the design and development of the product.

Nontechnical. Nontechnical risks are personnel, organizational, and process risks which are faced by the project manager. Some would say nontechnical risks are the most endemic since they stem from individual and workforce performance.

External. External risk is risk generated by the environment and the market, and can be anticipated through environmental scanning and strategic planning.

Predictable. Predictable uncertainty becomes risk because it can be anticipated, dimensioned, and mitigated.

Unpredictable. Unpredictable risk is uncertainty that cannot be anticipated or managed.

Legal. Legal risk is the probability that a project will generate legal action focused on the deliverable or proprietary information.

Risk event. A risk event is the triggering action, milestone, or task output that generates the risk and evidences that an anticipated risk is occurring.

Probability of occurrence. The probability of occurrence is estimated by the project manager, or rank ordered in terms of ranges of probability, such as 25, 50, and 75 percent.

Potential consequences. Consequences of risk are captured in the risk matrix not only in terms of impacts on schedule, cost, and quality, but also in terms of external consequences, such as change in company competency or market share.

Impact assessment techniques. Impacts are estimated by tools, which develop scenarios of future impacts through use of expert judgment and sometimes simulations.

Sensitivity analysis. This analysis looks at how project outcome is sensitive to various changes in schedule, cost, and quality, usually by trial and error.

Expected value. Expected value is the value of one path of decisions versus another, and is calculated by estimating the probabilities of various decisions and the profit margins outcomes of those decisions, then multiplying them and comparing various decision paths.

Decision tree. A *decision tree* is a tree diagram that outlines the various alternative paths at key points when the project manager must decide between contrasting decisions.

More Background

It is traditional to address project management in terms of time, schedule, and cost, and to focus on the Gantt chart as the key planning and control tool. Yet this emphasis on managing and sequencing tasks does not do justice to project management as a decision process. Project decisions are made (or not made) regularly on the basis of cost, risk, procurement and contracting, and other issues that can have major impacts on project success but which are not addressed in the traditional Gantt chart. The critical success factor for a project can often be the timing and nature of decisions made on the basis of risk and uncertainty and cost. These decision points in a project can figure importantly in the success of the project because they determine how risks and costs are handled and they narrow options.

Further, the timing of key decisions sets the conditions for what tasks are critical and what tasks become redundant as a project unfolds. But traditional project tools do not serve project managers well in this area, especially in flagging key decisions in terms of the project schedule and showing impacts as they are made. The risk process as described in the PMI *PMBOK* stresses risk planning, identification, assessment, and response. But this model does not help project managers anticipate key decisions and assess the costs and benefits (or expected value) of making one decision or another when key tradeoffs have to be made.

Project Tools

The combination of three analytic tools helps to illustrate the point: (1) decision tree analysis, (2) expected value, and (3) the traditional Gantt chart.

Decision tree theory

The decision tree helps to identify key decisions and evaluate expected value of key decisions based on the analysis of risk and costs associated with each branch.

There are two components to the decision tree—a decision and a risk or uncertainty. The decision is shown as a box with one arrow for each option available in the decision. The risk or uncertainty is represented by a circle and an arrow for each state of risk. The arrows for risk must contain the outcome in dollars if that state occurs, along with the probability of it occurring.

Note that the sum of all probabilities around the risk circle must add to 1.0. Therefore, the states must represent all possible conditions. These conditions are strung together to give a picture of the decision to be made. With the addition of a method for making a decision using the decision tree called the *expected value*, we can make our decision and have a method for presenting the results in a consistent manner.

An illustration of a decision tree analysis of risk

Pat is a project manager with a local contractor who has submitted a proposal to install a telephone trunk line between Macon central station and Kathlyne, GA. Pat has just received an option on 1000 acres of right-or-way property at $100/acre. If Pat purchases the option and the project is not selected, there is a 60 percent chance that she can sell the property at what it was brought for; otherwise she believes that there is a 60 percent chance she can sell it at $90/acre. Pat has another option to wait until the project is awarded to purchase the property. However, there is a 20 percent chance that the property will increase to $120/acre. Pat feels that there is a 60 percent chance that the company will be awarded the project. The original proposal, based on $100/acre, netted the company a profit of $100,000.

In building the decision tree for this problem we need to establish what decisions need to be made. In this case, the decision is to either “buy now” or “buy later.” Then we need to establish what the uncertainties are in the problem. In this problem, there are two uncertainties; the first is the same no matter which option we choose. That is, whether we win the contract. There is a 60 percent chance that we will win the contract. Therefore, there is a 40 percent chance we will not. The second uncertainty is different depending on the option being

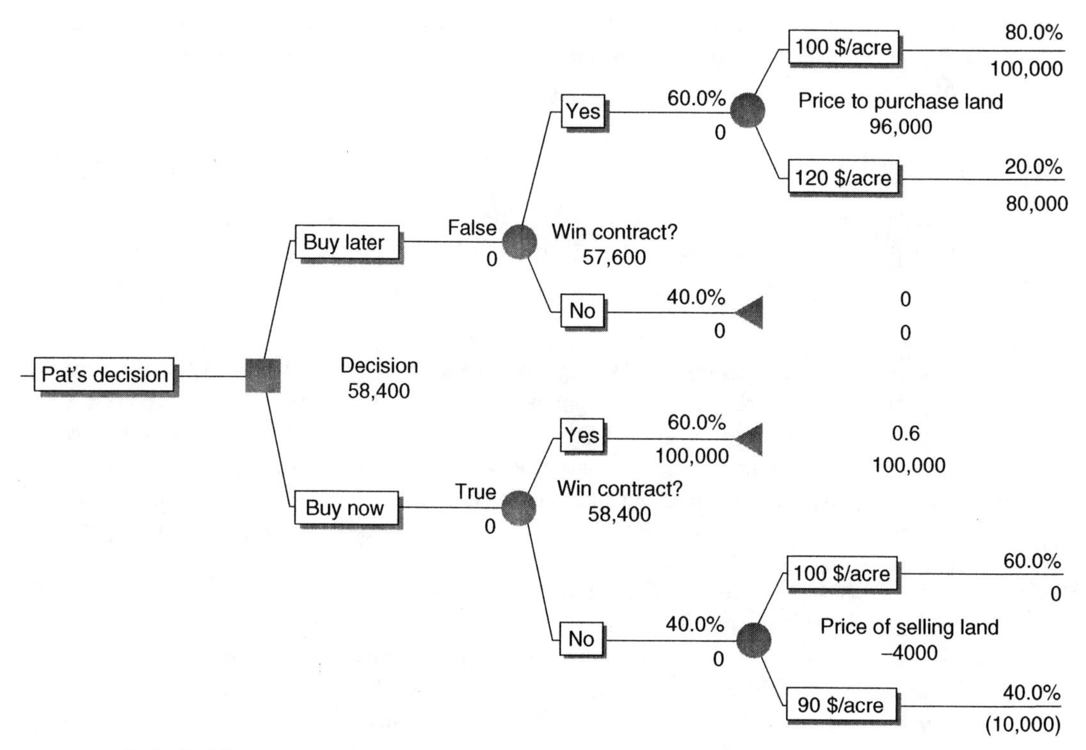

Figure 3.5 Pat's decision tree.

evaluated. That is, if we buy now and lose the contract, there is an uncertainty as to the price we can sell the land for to recoup our expenses. If we do not buy the land and win the contract, then we have to purchase the land. There is an uncertainty as to the price we can buy the land for at this point. The answer is calculated in the same manner above using expected value. When you have uncertainties that are chained together the expected value starts at the right-hand side and works left. You calculate the expected value around each uncertainty. The resulting expected value replaces the entire uncertainty in the next uncertainty. The resulting decision tree of Pat's decision process is shown in Fig. 3.5.

Using the decision tree, we start with the upper decision alternative of "buy later." At the left-hand side we see that the expected value of *purchase price* of land is 100,000(0.8) + 80,000(0.2) or $96,000. The entire uncertainty can be replaced by $96,000. Then the next uncertainty expected value can be determined as 96,000 (0.60) + 0(.4) or $57,600.

Following the same procedure, we can calculate the lower alternative of "buy now" with a result of $58,400. The tree can now be collapsed to the following decision:

We choose between an average profit of $57,600 for "buy later" or $58,400 for "buy now." Based on this, we would like to get as much profit as possible, so we would choose "buy now."

Issues

Simulation. Simulations create mathematical models using key factors in a future scenario and calculate outcomes based on various assumptions and input values. For the estimation of risk and impacts, simulations are useful when the project complexity is such that the interaction and impact of many factors cannot be handled manually.

Expert judgment. Expert judgment is gathered through Delphi techniques and focus groups who brainstorm to express their insights on a given risk problem.

Response planning

Prioritization. In prioritizing risks, the project manager develops a sense of where response planning is most needed. Risks are ranked in the risk matrix and assessed, and then the level of response planning and contingency is aligned with the relative importance of the risk.

Amount at stake. The amount at stake is actually a calculation of the expected value of a given risk decision; it is the dollar loss should a risk not be controlled, or should an opportunity not be exploited.

Response strategies

Avoidance. Avoidance is a viable approach, neglecting the risk, which sometimes "goes away."

Transference. Transference moves the risk to another party, as in insurance or contracting.

Mitigation. Mitigation planning takes the risk on squarely and provides the project manager with a direct corrective action.

Acceptance. Acceptance is the approach that essentially plans for the risk to occur and plans for covering the cost and schedule impacts.

Monitoring and control

Risk response audits. Risk response audits are used to look back at how a given project or project risk has been handled. A project audit looks more broadly at how the project was managed across the board.

Periodic project risk reviews. Risk reviews are incorporated into project reviews, not in separate sessions. The project review agenda includes risk analysis and planning data and poses risk decisions to be made.

Earned value analysis. Earned value monitors schedule and cost variance as indicators of risk impacts.

Communicating risks. Risks are addressed in business and project reports, which anticipate risks, explain contingencies, and pose decisions to be made which may shape the risk impact.

Project Control Systems

Cost/schedule integration

Information flow and ties to WBS. Baseline schedules are developed from the WBS and schedule and cost data are related to particular tasks. Since costs are directly associated with tasks, each task can be assessed in terms of cost and schedule impacts.

Network planning and schedule development. Scheduling is the most important function of project managers and risk determines the amount of "buffer" that is withheld by the project manager based on the probabilities of risk occurrence and contingency action.

Project cash flows and commitments. Cash flows committed out beyond customer agreements or contracts represent separate risks, thus cash flows are aligned with work performed and scope boundaries.

Reporting responsibilities. Project managers report risk and cost information to top management, stakeholders and customers; functional managers report technical and technology risks and costs to top management.

Issues

Cost, schedule, performance/quality tradeoffs. Decisions which trade off schedule, cost, and quality/performance factors can be guided by the risk impacts of each option. In other words, when a project manager decides to delay a task outcome because of quality considerations, the determining factor might be the risk that the quality outcome cannot be achieved even with more time provided by a task delay.

Integrated change management. Change management is the process of accommodating to a change request by reviewing all internal and external impacts, reviewing the risks of change, and integrating the change across all appropriate interfaces.

Corrective action

Crashing. Sometimes the project manager must direct new resources to a project to make up for unanticipated impacts of risk events, thus crashing the project. The risk of quality impacts of crashing can be found in the tendency for "haste to make waste."

Phasing of deliverables. Phasing deliverables in a different way may help to relieve the tensions of a project task delay or failure.

Modification of scope, schedule, budget. Sometimes a scope or schedule must be altered because of schedule or cost variance, but there are inherent risks in doing so.

Chapter

4

Demystifying Risk: Using the PMI *PMBOK**

Demystifying Risk—*PMBOK*

It is not only important to know the PMI *PMBOK* guide on risk management but also to simplify the process to match it with reality. The PMI *PMBOK* defines risk management as the "systematic process of identifying, analyzing, and responding to project risk." The concept aims at maximizing the probability and consequences of positive events and minimizing the probability and consequences of adverse events to project objectives.

It is important to see the *PMBOK* as a guide, not a manual. While the following discussion is guided by the *PMBOK* (2000 edition), the reader will quickly see that each section is framed by a *reality check*—the author's personal and professional view of best practices in the real world of *faster, better, and cheaper.* The discussion presupposes a separate risk management planning process, which serves as an ideal. In practice, risk planning occurs as *an integral part of project planning.*

The *PMBOK* framework is very useful as a process guideline, but in view of developments in enterprise project management, multiproject environments, project and portfolio selection, and the new appreciation for the interfaces of project management with other organizational systems, the *PMBOK* is only a start. Table 4.1 contrasts the current *PMBOK* on risk with future needs.

Process. We are learning that while process focus is important, it misses the opportunity to integrate risk into current business and project planning and management actions. Process focus is useful for ensuring quality and discipline, but has limitations in practical work settings.

---

* *A Guide to the Project Management Body of Knowledge,* 2000 edition, Project Management Institute.

TABLE 4.1 Comparison of Current *PMBOK* Standards on Risk with Future Needs

Process focused	Management focused
Separate procedure	Integrated into project planning and control
Single project oriented	Portfolio, multiproject oriented
Emphasis on quantitative	Emphasis on qualitative and professional judgment
Focused on methods and procedures, not people	Focus on training and enabling people to manage risk
Assumes inputs to process exist	Realistic picture of inputs
Not related to cost	Integrates risk and cost
Not related to quality	Integrates risk and quality
Ignores business wide risk	Initial focus on business risk strategy
Does not incorporate contingency into planning	Integrates risk into the scheduling process
Ignores risk as opportunity	Connects risk control to opportunity

Separate process. The propensity to breakdown the project planning and control process into components misses the actual dynamic in real organizations where everything goes on all the time. Risk has not proven to be useful as a separate process, but rather effective only if integrated.

Single project. Risk is no longer looked at as a single project issue; most risk is associated with broader issues, such as the business itself and other projects in the company portfolio.

Quantitative. Overemphasis on quantitative tools and mathematical models suggests risk management as a science rather than art. Most risk management actions are full of judgment and margins of error, which make quantitative tools ineffective and intimidating.

Focus on methods, not people. The risk process is essentially a thought process, a way or style of management that is ingrained in the way people work and solve project problems.

Overemphasis on methods and undervaluing of the human element limit the application of the current *PMBOK.*

Assumes inputs. The assumption in any process focus is that the inputs will be there, but in many cases the necessary inputs are not there because the whole concept of separate risk management process is flawed when looked at in practical terms.

Unrelated to cost. Risk and cost are inextricably intertwined and therefore it is difficult to look at risk without looking at the costs of control and costs of impact.

Unrelated to quality. Risk and quality are also inextricably connected. Since risk has an impact on quality, many risks are associated with feasibility, not schedule—some projects may not be capable of meeting customer standards and specifications in the first place.

Ignores business risk. Project risk is first identified and managed at the businesswide level, through strategic and business planning.

Separate contingency planning. The current *PMBOK* describes contingency planning as a separate process, but in order to be effective contingency actions need to be incorporated into baseline schedules and budgets. The project manager ensures that the schedule has buffers and contingency tasks built in.

Ignores risk as opportunity. The other side of risk is opportunity—the ability to control risk creates opportunity because the competition cannot. Thus any project aimed at capturing a market share is designed to create opportunity.

Table 4.2 provides an overview of the current *PMBOK* risk processes.

- *Risk management planning.* Deciding how to approach and plan the risk management activities for a project
- *Risk identification.* Determining which risks might affect the project and documenting their characteristics
- *Qualitative risk analysis.* Performing a qualitative analysis of risks and conditions to prioritize their effects on project objectives

TABLE 4.2 *PMBOK* Risk Management Processes

11.1 Risk management planning	11.2 Risk identification	11.3 Qualitative risk analysis
.1 Inputs .1 Project charter .2 Organization's risk management policies .3 Defined roles and responsibilities .4 Stakeholder risk tolerances .5 Template for the organization's risk management plan .6 Work breakdown structure	**.1 Inputs** .1 Risk management plan .2 Project planning outputs .3 Risk categories .4 Historical information	**.1 Inputs** .1 Risk management plan .2 Identified risks .3 Project status .4 Project type .5 Data precision .6 Scales of probability and impacts .7 Assumptions
.2 Tools and techniques .1 Planned meetings	**.2 Tools and techniques** .1 Documentation reviews .2 Information gathering techniques .3 Checklists .4 Assumptions analysis .5 Diagramming techniques	**.2 Tools and techniques** .1 Risk probability and impact .2 Probability/impact risk rating matrix .3 Project assumptions .4 Data precision ranking
.3 Outputs .1 Risk management plan	**.3 Outputs** .1 Risks .2 Triggers .3 Inputs to other processes	**.3 Outputs** .1 Overall risk ranking .2 List of prioritized tasks .3 List of risks for additional analysis and management .4 Trends in qualitative risk analysis results

(*Continued*)

TABLE 4.2 ***PMBOK*** **Risk Management Processes (*Continued*)**

11.4 Quantitative risk analysis	11.5 Risk response planning	11.6 Risk monitoring and control
.1 Inputs .1 Risk management plan .2 Identified risks .3 List of prioritized risks .4 List of risks for additional risk analysis and management .5 Historical information .6 Expert judgment .7 Other planning outputs	**.1 Inputs** .1 Risk management plan .2 List of prioritized tasks .3 Risk ranking of the project .4 Prioritized list of quantified risks .5 Probabilistic analysis of the project .6 Probability of achieving the cost and time objectives .7 List of potential responses .8 Risk thresholds .9 Risk thresholds .10 Common risk causes .11 Trends in quantitative and qualitative analysis results	**.1 Inputs** .1 Risk management plan .2 Risk response plan .3 Project communication .4 Additional risk identification and analysis .5 Scope changes
.2 Tools and techniques .1 Interviewing .2 Sensitivity analysis .3 Decision analysis .4 Simulation	**.2 Tools and techniques** .1 Avoidance .2 Transference .3 Mitigation .4 Acceptance	**.2 Tools and techniques** .1 Project risk response audits .2 Periodic project risk reviews .3 Earned value analysis .4 Technical performance measurement .5 Additional risk response planning
.3 Outputs .1 Prioritized list of quantified risks .2 Probabilistic analysis of the project .3 Probability of achieving the cost and time objectives .4 Trends in quantitative risk analysis results	**.3 Outputs** .1 Risk response plan .2 Residual risks .3 Secondary risks .4 Contractual risks .5 Contingency reserve amounts needed .6 Inputs to other processes .7 Inputs to a revised project plan	**.3 Outputs** .1 Workaround plans .2 Corrective action .3 Project change requests .4 Updates to the risk response plan .5 Risk database .6 Updates to risk identification checklists

- *Quantitative risk analysis.* Measuring the probability and consequences of risks and estimating their implications for project objectives
- *Risk response planning.* Developing procedures and techniques to enhance opportunities and reduce threats to the project's objectives
- *Risk monitoring and control.* Monitoring residual risks, identifying new risks, executing risk reduction plans, and evaluating their effectiveness throughout the project life cycle.

These processes interact with each other and with the processes in the other *PMBOK* knowledge areas. The way they interact is key to integrating risk management with the basic project planning and control process. Here are the salient points of integration.

Risk Management Planning

Risk management planning for a particular project is inextricably connected to how the organization prepares for dealing with risk and uncertainty in its business development and strategic planning, in its information technology investments and management of network communications, and in its organizational structure. No project manager faces risk alone—it is a company-wide issue and it is quite likely that there are data and information on project risk in the company files.

A company prepares for risk in projects by assessing the overall risk in strategic planning. Then, in its development of a program of projects, or portfolio, the company assigns risk to a general program or product line as part of its decision to proceed. Templates for risk identification and assessment are available.

A project management office is sometimes available to support risk management by providing for information templates, project review data and agendas, research findings, and historic information on various risk subjects.

To address risk effectively, there needs to be an information technology capacity in the company to ensure that risk documentation and tracking can be achieved within the network and the software assets available. This means a way to organize risk data, for team members to communicate risk information quickly, and for project managers to present risk data in acceptable formats.

Finally, if the company is not organized to address projects in some kind of project structure, project risk will get the same kind of attention other project issues, such as cost and schedule, get—very little. In a matrix structure, for instance, risk is addressed in the functional department in terms of processes and equipment to address risk through testing and monitoring technical processes. At the same time, the project manager is attuned to risk when schedules are delayed because of events or developments grounded in risk.

This does not need to complicate the up-front risk process. *Risk is still a relatively straightforward concept of planning and controlling for things that could go wrong.* That risk management planning is tied to business planning and strategy need not suggest that this linkage complicates the achievement of a good risk management process. In fact, it makes project risk management easier in the sense that business planning itself provides a precursor to project risk. If I am in the avionics business, and I design and produce avionics equipment for business jets, and my overall business plan identifies major threats to profitability and success, such as the global availability of cheap LCD monitors or the pending change in air traffic regulations affecting avionics, then my individual projects begin with a major challenge in those two areas. As a project manager in that scenario, I enter the project arena with a *built-in*, up-front view that

I must keep my eye on both these factors and plan and schedule contingency and risk mitigation actions as part of my project plan.

The *PMBOK* process view emphasizes the inputs and outputs of the risk process. This is a useful perspective on risk even though it does not really deal with integration of risk management into the project planning and control process. The *PMBOK* is based on the highest level of "maturity" in an organization, a scale that is embodied in the PMI maturity model. Thus the *PMBOK* process is idealized in a *mature organization*, and rarely found in all its dazzling performance dimensions in a normal business environment.

Inputs to risk management planning

Project charter. The project charter is an ideal project planning document that includes the business need and product description. In reality, this document is often neglected because the content for the business need is still in conceptual stages. And since at this point the deliverable is often undefined and unspecified, the product description is not at the level of detail of a configuration management document. But the deliverable is defined in performance terms at the scale possible, given the understanding of what is being designed and built.

Organization's risk management policies. If the business has a set of risk management policies and procedures, they would be used in risk planning. However, many businesses do not have such policies, nor will they; rather they apply risk tools as an implicit part of the project planning process. Approaching this process input with a healthy skepticism one can see the value of writing down policies and procedures, but in today's fast moving companies this is rarely done. The point is that a nimble midsized business of today that has articulated its approach to risk would expect each employee and certainly its management to embody that approach in the basic project management process—*without a bureaucratic statement of top-down policy.*

Defined roles and responsibilities. One would hope that the basic roles and functions of the project manager and functional manager is documented, but again this is often left undefined in order to allow a natural process of negotiating and working out roles between functional quality and project delivery interests.

Stakeholder risk tolerances. The *PMBOK* is not very helpful in illustrating this input. Stakeholder risk tolerances are evidenced in the modern business as "world views" of certain important people in the process, such as sponsors, customers, investors, top management, and regulators. A risk tolerance for an electronic instrument might be framed as a technical tolerance (e.g., mean-time-between-failures), or a performance tolerance (e.g., must perform in below zero temperatures), or a drop-dead limit (e.g., we investors will not proceed with this project if by this time next year there is no first article production unit because of the anticipated rapid change in market conditions). Tolerances are often grounded in expectations, thus it is important for a project manager to *see* such tolerances and evaluate their intensity early in the project.

Work breakdown structure. Certainly, a WBS is necessary as a basis for identifying risk, but the WBS must be comprehensive and show all tasks before the risk identification process can really be effective. If the WBS misses some important work that is highly subject to failure then the WBS is not a good input to risk management planning.

Issues not addressed in *PMBOK*

Some risk issues are not addressed squarely in the *PMBOK* but are important in gaining a full understanding.

Building a risk-based organizational culture. Building an organization that protects itself from project level risk and uncertainty through good organizational planning and management requires strong leadership. As a project manager you have to first *feel* that you are expected by your management to anticipate and deal with risk and that you will be supported in taking the time and investing the cost involved in building a good risk planning and control process. Risk management starts at the top leadership level with clearly articulated vision and mission that incorporates an uncompromising commitment to quality and excellence. In so doing, the leadership commits also to a risk management process to reduce the probability of failure and to promote total quality in product design and production.

Program and portfolio management. The management of risk in a multiproject environment, and the role of risk in selecting and maintaining a portfolio of balanced projects are not really addressed in the *PMBOK*. Yet the selection of the *right* projects for the project pipeline inherently involves risk management. Projects with high risks must be identified before they enter the approved list simply because a portfolio of high-risk projects endangers the long-term growth and profitability of the enterprise.

Interface management. Good risk management is dependent on the availability of effective support and interface services to project managers. Risk cannot be seen simply as a project management issue; it is an accounting and cost issue, a procurement issue, and an information technology issue. Cost data must be available to project managers to assess cost impacts; contract officers must be attuned to risk and risk sharing issues in managing contractors and supply vendors; information technology administrators should see the need for web-based, easily available risk matrix templates and calculations software.

Risk and cost integration. For some reason, the relationship of cost and risk is lost in the daily routine of project managers, yet it is in cost and "expected value" that future impacts of risk decisions can be made. Contingency plans add to project cost estimates and when there are clear decisions that must be made and crossroad tradeoffs to be decided, there must be a support system available on making those decisions.

Tools and techniques for risk management planning

Planning meetings. Planning meetings are important and can be fruitful, but they can also be a complete waste of time unless they are focused and deliver results. Setting agendas, facilitating meetings, and writing good follow-up notes are all useful tools.

The *PMBOK* treatment of risk management planning does not cover some of the most important risk management planning tools, namely:

1. *Business plan.* The grasp of the business plan helps a prospective project manager get an early start in project risk management planning. Such a plan, or a business strategic plan, will provide strategic information, e.g., SWOT (Strength, Weakness, Opportunity, and Threat) data and information.
2. *WBS.* The WBS is an early indication of the potential risks in any project, and planning for project risk requires at least an outline of the WBS to see the basic components of work involved. Risk management planning requires the project manager to anticipate how risks will be handled by looking at the WBS, or building one.
3. *Information and network systems.* A major risk management planning tool is the company network and information sharing system and data already available on similar projects in the past.

Outputs from risk management planning

Risk management plan. The plan outlines the approach to how risks will be handled in the project. Frankly, many companies do not need a separate risk management plan, but they should integrate risk information into the project scope, WBS (definitions), schedule, and budget. As a threshold, if the project under consideration is more than 300 tasks and budgeted at more than $5 million, a separate risk management plan is called for.

The plan includes:

1. *Methodology.* It is not clear from the *PMBOK* what the "methodology" of the plan is, but it appears that the concept says you should have a methodology. For instance, an electronic instrument production firm should use safety and reliability tools, such as mean-time-between-failures, to test its prototypes to avoid the risk of performance variation and failure.
2. *Roles and responsibilities.* Here the plan addresses who is responsible for what in a functional and project management context. Here would be where the role of a program management office would be described.

3. *Budgeting.* This is a budgeting exercise to estimate the cost of risk management. Frankly, it would be more important to spend time analyzing the cost of the risk event. The cost of risk management is a program management cost category, as in quality assurance and project review.
4. *Timing.* This addresses when various risk management actions will be taken in the project schedule, such as risk analyses, preparation of contingency plans, and response plans.
5. *Scoring and interpretation.* This portion of the risk management plan addresses tools, such as the weighted scoring model (aligns projects in project selection with business strategy, places priority weights on various strategic objectives, and scores each project against the strategic objectives) and cash flow, rate of return, and net present value.
6. *Thresholds.* Thresholds address the criteria—rules of thumb—for acting on risks or to reduce risks, such as deploy preventative contingency and response plans for risks which could delay a project by more than 10 percent of the total project duration unless the risk is reduced in the first quarter of the project.
7. *Reporting formats.* This provides guidance to the project manager on the formats, e.g., email, MS Project team reporting, hardcopy spreadsheets, for project reports to stakeholders.
8. *Tracking.* This provides guidance on what risks will be tracked and how, such as the acquisition of microchips for an electronic instrument will be tracked with the contractor on the basis of earned value.

Risk Identification

Risk identification should be part of the project planning process, not separated from it. Risks are identified in the development of the WBS and in estimating durations and resource needs.

Inputs to risk identification

Risk management plan. To the extent that a risk management plan is produced, it is an important input to identifying risks. But since most companies will start the process of handling risk with the WBS and the task list, the identification of risk usually starts in earnest in the review and final production of the generic WBS. It is here that the project manager *reviews every task for its potential for failure.* Identifying risk involves a lot of discussions with team members and stakeholders. For instance, if a technical task in the WBS, say achieving a given mean-time-between-failures in a product component, "sticks out" initially because of the challenges of completing it then it is the subject of much discussion and contingency planning early in the project.

Project planning outputs

Project charter. If there is a charter, the charter should be helpful in identifying risks and confirming the judgment of the project manager on where the project vulnerabilities are.

WBS. The WBS is the basic source of risk identification activity since it embodies all the work of the project, or should. The WBS should be four levels down to give enough detail to the project profile to *see* project risks.

Product description. The product description will be embodied in a configuration management document, or in some document/drawing/specification that defines the product from a performance and component perspective. This will occur in the design phase at some point when the deliverable has been fully fleshed out.

Schedule and cost estimates. The schedule and project budget (part of the schedule in MS Project) will be a good source to confirm risks, but first the project manager must prepare a risk matrix as described earlier. The risk matrix identifies the task, the task risk description, and the impact (schedule, cost, and the like).

Resource plan. While there is typically no formal resource plan, there will be a sense of the personnel, equipment, capital, space, and technology needs of the project. Ideally, this resource plan is in one place, for example, in MS Project and/or in a planning document of some sort. But at this point, if *all* the resources needed for the project are not clear, it is not critical. What *is* needed here is a clear idea of the "bottleneck" resources—those resource issues that could represent a barrier to achievement of the project. These bottleneck resources might be a critical software engineer who is already spread too thinly in current projects, a piece of testing equipment that is critical to meeting quality control thresholds, or a work space or station that is being shared with other projects.

Here also is the beginning of the application of the theory of constraints. Simply put, the theory states that the focus of attention in planning and control should not be the whole project and all its tasks, but rather the one or two major resource constraints. The project manager protects against the risks inherent in these resources by tapping time and cost from the original estimates and withholding them for allocation when they are needed.

Assumption and constraint lists. The assumption list is a convenient way of indicating the controlling assumptions, e.g., the assumption that a sole source contract with a foreign supplier for a key product component will last through the project life cycle. In practice, these assumptions are well known by the project team and stakeholders, but it pays to document them and revisit them in project review sessions and to treat them as risks with contingency plans.

Risk categories

Technical. Technical risks have to do with product, process, or "technique" issues involved with designing and producing the deliverable.

Project management risks. Project management risks address the things that can go wrong with the project planning and control process and with expected support services from the information technology source and a project management office, if the organization has one.

Organizational risks. These are the "soft" issues that a project faces, which have to do with organizational behavior and dynamics, e.g., conflicts, scarce resources, personnel performance problems, and company-wide crises, such as lack of financing or a downturn in share value. A key organizational risk is the lack of top management support, which will be evidenced with the neglect of the project in the company "head-shed."

External risks. External risks are the business and global risks inherent in any business, such as economic downturns, trade difficulties affecting the deliverable, and the impacts of communication in multinational companies.

Historical information. Historical information would include past project documents, "lessons learned" reports, and industry information available on the competition, on demand and market issues, and on the company's performance on similar projects in the past.

Project files. Project files would be available from the company's file system, but in practice project managers rarely look back even though it would make sense to do so.

Published information. This would include manuals, articles, and technical publications on the deliverable.

Tools and techniques for risk identification

Documentation reviews. The key document in risk identification is the WBS. Other documents would include past project reviews and similar product performance information.

Information gathering techniques. These would include web-based information, electronic files with product information, and Internet research.

Brainstorming. Brainstorming is simply having meetings with key people who know something about the project and generating ideas and options without judging them. Brainstorming generates ideas and does not filter them.

Delphi technique. Delphi is brainstorming with key experts who go through a systematic process of providing their views, reviewing each other's ideas, and coming up with a scenario based on the integration of their views.

Interviewing. Interviewing key stakeholders, past project managers, and task managers helps to uncover "subtle" information that has not been documented.

Strengths, weaknesses, opportunities, threats. Here we look back at the business planning processes for strategic analyses, especially including threats and opportunities.

Checklists. Checklists are typically prepared by a documentation specialist for various project and product documents. Checklists often key into potential failure points in past projects and thus are very useful in identifying risks.

Assumptions analysis. The key source of assumptions is rarely captured in one document, but the concept of focusing on assumptions is important. A project management office is typically in charge of documenting assumptions.

Diagramming techniques. Flow charts and diagrams, such as decision trees, are useful in identifying the various options and decisions, including expected value.

Cause and effect. "Root causes" of risks can be identified through fish-bone diagrams and other such meeting techniques.

Influence diagrams. Diagrams that indicate cause and effect, and influence of key factors, can help in identifying risks.

Outputs from risk identification

Risks. The output of risk identification is a better sense of risks over time. In truth, the clarity of risks increases as time goes on in the project life cycle.

Triggers. Triggers of risk include those indicators or signals of risk events that become clearer in the risk identification process. For instance, in a contract negotiation in the outsourcing process, a contractor refuses to sign the contract because of a schedule requirement for delivery of a key supply or piece of equipment before a key project milestone.

Qualitative Risk Analysis

PMBOK separates the risk analysis process into two parts: (1) qualitative and (2) quantitative. Qualitative connotes a better description of the risk, its dimensions and its characteristics; quantitative involves getting a finer cut on risk by applying mathematical and other quantitative tools. In practice, companies rarely split the two. The point of risk analysis is to drill down on potentially high-risk tasks to get a more detailed picture of their impacts.

Inputs to qualitative risk analysis

Risk management plan. The risk plan again is useful.

Identified risks. A completed risk matrix is required before risk analysis proceeds.

Project status. The timing of risk analysis is important because the risk impacts, particularly in terms of schedule and budget, will change depending on when they are analyzed. The later in the project cycle, the clearer the impacts will be.

Project type. It is important to *dimension* the risk here; a project producing an electronic instrument for a sophisticated aeronautic application will get a different look than a project to construct a standard building.

Data precision. The accuracy and precision of data are important; if you know that a given reliability test has a proven failure rate then the results must be tempered accordingly.

Scales of probability. Probability is a subjective judgment unless the product is tested many times to develop a statistical mean. In most projects, the probability of a risk occurring is the result of thinking through how many times this kind of risk has occurred in past similar projects, combined with "gut" judgment of the project manager and key stakeholders.

Assumptions. Again, the assumptions are rarely listed, but they are apparent in the analysis process.

Tools and techniques for qualitative risk analysis

Risk probability and impact. Using the risk matrix, the project manager has already identified risks and will assign probabilities to all high-impact risks. For most risks, these probabilities are subjective and simply communicate a sense of confidence that the project manager has about the risk in question. Generally, probabilities should be stated in three forms, 25 percent, 50 percent, or 75 percent, suggesting little change of the risk occurring, substantial chance, or high chance, respectively.

Probability/impact risk rating matrix. The rating or risk matrix now is fine tuned in the analysis process, with more data and more attention.

Project assumptions testing. Testing assumptions involves taking time to review key assumptions and confirming that the assumption is right and that the probability assigned is in the right ballpark.

Data precision ranking. For most projects, this step is not useful. If the product is highly complex and must meet detailed performance specifications then the data precision ranking indicates how precise the test data are compared to a common standard.

Outputs from qualitative risk analysis

Overall risk ranking for the project. Once the qualitative process is finished, two rankings are produced: (1) how the project's overall risk is ranked compared to others (may be completed in comparing and selecting a portfolio of projects) and (2) how individual risks rank within the project, usually limiting the list to five or less.

Again, this process is related closely to the theory of constraints. A project typically faces only a few major bottlenecks or risks, and it is the job of the project manager to accurately uncover those few critical risks during qualitative analysis.

List of prioritized risks. The list of prioritized risks is incorporated in a project report to stakeholders along with supportive information including the risk matrix and contingency plans.

List of risks for additional analysis and management. A residual list includes other risks that could turn out to be more important than they appear.

Trends in qualitative risk analysis results. Some review of the credibility of the process will uncover past analyses and how their results played out in real terms.

Quantitative Risk Analysis

Inputs to quantitative analysis

All the inputs to qualitative analysis are relevant here, plus expert judgment. Expert judgment comes from technical experts who have knowledge at the technical level on the risk in question. For instance, here is where a project manager brings in a safety and reliability expert contractor to advise on variability thresholds in testing prototype parts.

Tools and techniques for quantitative risk analysis

Interviewing. Again, interviews are useful with key experts on defined task risks.

Sensitivity analysis. Sensitivity analysis determines how much project outcomes, e.g., schedule, budget, and quality are sensitive to particular risks. For instance, it may turn out that although a risk is only medium probability and impact, it could alter the final deliverable in a measurable way in terms of quality control.

Decision tree analysis. Decision tree analysis aims to uncover expected value of taking one path or another when a project "crossroad" decision must be made. In the case already discussed, a decision on whether to purchase land in anticipation of winning a contract brings with it a set of expected values for going one way or the other.

Simulation. Simulations are mathematical representations of scenarios involving key project risks. This might include an equation that specifies what happens to an information technology scheme when its capacity is challenged by demand, e.g., shut downs, performance failures, and the like.

Outputs from quantitative risk analysis

Prioritized list of quantitative risks. Quantitative analysis usually places more content on the already produced list of risks from qualitative analysis. New data are presented on high risks from the analysis.

Probabilistic analysis of the project. Here the probabilities are worked to a finer level of detail based on more analysis. A probability set at 25 percent in the qualitative phase might be fine tuned here to, say, 38 percent with more input from intensive analysis of past projects and simulations, and perhaps some experiments.

Probability of achieving the cost and time objective. A final probability is determined for meeting the project schedule and budget goals.

Risk Response Planning

Appropriately, the *PMBOK* places high priority on a response plan that mitigates risk. That response plan is grounded in the contingency plans already developed in preparing the risk matrix. The key point about response planning is to outline corrective action and to incorporate those actions in the baseline schedule so that they will not later be considered add-ons or changes to the project. They might later be identified with the "pessimistic" estimates of durations in the PERT analysis in MS Project.

Inputs to risk response planning

Risk thresholds. Risk thresholds help to identify acceptable ranges for risks to occur without deploying contingencies, e.g., if this work is not done in the estimated time then we will give the team 3 more weeks before we act since the task is not on the critical path. They come from customers, stakeholders, and technical experts.

Risk owners. Risk owners are those stakeholders who are accountable for acting on risks, or at least reporting on risk activity. If there is no designated risk owner, a risk could be unattended long after it is identified because of the tendency to avoid controversy and accountability for risks that cannot easily be controlled.

Common risk causes. Common to all industries are a set of common risk causes, such as government regulatory change, bad marketing information, faulty safety and reliability equipment, and lack of proven competencies in particular personnel categories.

Tools and techniques for risk response planning

Avoidance. One approach to a risk is to avoid it and hope that it goes away. Sometimes it does.

Transference. Transference involves turning a risk over to a risk owner, e.g., assigning a contractor the job of responding and providing incentives for risk reduction.

Mitigation. This is the corrective action option leading to deployment of a contingency plan.

Acceptance. Sometimes it pays to accept a risk and deal with it directly rather than transferring it, avoiding it, or mitigating it. "Living with" the risk means plugging in schedule and budget reserves assuming that the risk cannot be controlled and working around it.

Outputs from risk response planning

Risk response plan. Although a formal, written risk response plan is not always feasible or wise because of the cost and effort involved, it is the *learning and thinking* process that follows from good risk response planning that gives it value. Once through with the process of defining risks, planning to respond, and folding the results into planning documents, such as the schedule and budget, a project manager "owns" those risk responses and has incorporated them into the plan. The lack of a separate, documented risk response plan is not a good indicator of the fact that risks have not been considered. It is more important that risks are incorporated in the project schedule and estimates made from "expected, optimistic, and pessimistic views," which are driven from risk analysis.

Residual risks. Residual risks are risks that continue to exist after corrective action. Sometimes residual risks are created in taking a corrective action that was not anticipated in the original project planning process.

Secondary risks. These are lower level risks that have less impact but which can grow in importance if neglected.

Contractual agreements. The type of contract used in outsourcing work involves some explicit assumptions about risk transference. For instance, a fixed price contract is superior to a cost reimbursable contract in transferring risk to a contractor to control a given risk in performing contracted project work.

Contingency reserve amounts needed. Sometimes a project must be protected with a reserve fund, or insurance program, so that the company is not financially exposed from a given risk even though it is mitigated.

Risk Monitoring and Control

The *PMBOK* places emphasis on monitoring risks and controlling them, but this process is again an integral part of the project review and control process.

Inputs to risk monitoring and control

In addition to the common inputs addressed above, the new inputs to monitoring are communication and scope change.

Project communication. Communication means exchanging information on anticipated risks so that people who have a stake in the project can assist in mitigation and can adjust their expectations for the project. Communication always helps to reduce the uncertainty and "surprise" factor in dealing with customers, clients, and stakeholders.

Scope changes. As the project is progressing and work is getting done, there may be new information uncovered in the risk management process that requires a change in the scope, schedule, budget, or quality standards in the project. Thus scope change is a logical outcome of monitoring and seeing risks impact the project.

Tools and techniques for risk monitoring and control

Project risk response audits. A project risk audit looks back at how effectively project management processes in general were handled and how well project risks were monitored and mitigated.

Periodic risk reviews. Risk review occurs in the normal project review process, not as a separate process. A standard project review agenda always includes a section on risks.

Earned value analysis. Schedule and cost variances always indicate the possibility of risks being at work if the project is not performing as planned. Corrective action when uncovering major earned value variances over 10 percent would include a review of risk contingencies and impacts.

Technical performance measurement. Technical issues could be at work in a project that is slipping or meeting insurmountable obstacles.

Outputs from risk monitoring and control

Workaround plans. The so-called workaround plan is a manifestation of the risk mitigation process. Workaround is a contingency that should be identified in preparing the risk matrix. Workaround sometimes takes advantage of innovative, creative options to overcome a project risk, and is often the result of "out-of-the-box" thinking that can be generated in a brainstorming session.

Corrective action. Corrective action is the action taken when the project is not performing according to plan—schedule, cost, and quality.

Project change requests. In monitoring, the need to fundamentally change a project scope or key deliverable occurs often, thus triggering a change request.

Updates to the risk response plan. Updates to the risk response plan result from lessons learned in the monitoring process.

Risk database. The risk database is a documentation of risk information that will be useful in corrective action and future projects.

Updates to risk identification checklists. Updates to risk identification checklists help keep tabs on best practices in risk mitigation as they are generated.

Summary of Risk Management Process

Identify and categorize risks

Identifying risks is not a science, it is an art. It does not require a sophisticated, mathematical exercise, although some measurement may be useful to dimension the risk.

Given a project description including project goals, work breakdown, schedule, and resource assignments, identify and categorize potential risks using various tools including risk assessment, brainstorming, peer review, and document review.

Here we develop the basic skill of reviewing a project scope, work breakdown, and schedule, and identifying risks from the tasks and processes. The reader goes through a risk matrix, which lists potential risks, defines them, categorizes them, estimates impact on various project performance outcomes, such as schedule, budget, and specification, and might include reference to a contingency plan—a plan to take an action if a "worst case" actually happens.

Assess risks

Assessing risks is part of the planning process, not a separate, quantitative, or qualitative exercise.

Given a project description and identified risks, apply various quantitative and qualitative analysis tools, including sensitivity analysis, to determine the effects, outcomes, and consequences of identified risks.

This is where we will apply tools and methods to assess impacts and consequences of various risks on project success. You can use a variety of tools. Sensitivity analysis tries to figure out how sensitive the project is to a given risk and there are a variety of approaches to sensitivity analysis.

In the real world, project managers don't do sophisticated probability analysis to get the probability of something happening to 88.54 percent. The margin of error is so large in trying to determine the probability that you are lucky to get in the ballpark. The real issue is saying, "Here is a potential risk. Is the probability of it happening 25 percent or 85 percent? If it's 85 percent, I better do something about it."

Risk assessment tools. Here we get into the application of risk assessment tools. Risk assessment is the process of analyzing project risks to learn more about them, to quantify probabilities when it makes sense, and to help decide what to do about them. We will learn about qualitative and quantitative assessment. Qualitative assessment focuses on describing risks, ranking them, and making sure they are fully understood. Quantitative assessment applies probability and other analytic tools, such as decision trees, to pin down the extent of risk and risk impacts.

We will apply some quantitative tools. While risk management is not primarily a quantitative exercise, it sometimes helps to quantify the risk and its impacts to get a better handle on it.

Risk assessment involves two basic functions: (1) qualitative—rank ordering risks so that you can decide which are the most important to tackle and (2) quantitative—doing more quantitative analysis to pin down risks so that risk response can be predicated on quantitative measures, such as the probability that something will happen and have the anticipated impact, and so that stakeholders can be assured that risks have been analyzed.

The goals of risk assessment are as follows:

1. To increase the understanding of the project—the more I know about risks, the more I know about the project
2. To rank-order risks in terms of severity and impact so that I know which to address first
3. To serve as the basis for identifying alternative approaches to response and risk management and integrating risk into the risk management process
4. To offset the normal tendency to be optimistic in project planning—the basic human trait of expecting the best will take over unless you weigh it against a risk analysis that forces you to identify impacts and worst case, pessimistic scenarios.

Risk qualification

Risk qualification involves listing risks and using a risk matrix to determine how risks should be categorized and ranked according to their impacts on schedule, quality, cost, and overall success of the project.

Risk quantification

The tools for risk assessment we will address are:

1. *Probability.* Probability analysis is the process of developing a measure of how likely a risk is to (a) occur and (b) create the impacts anticipated. We want to be able to say that not only is this risk liable to happen, but there is a 50 percent probability that it will happen. A 50 percent probability that it will happen combined with a 50 percent probability that if it happens it will have the impact identified, creates real urgency to respond, as opposed to a risk that has only a 20 percent probability of happening.

   Remember that the initial determination of the probability that a risk will occur is typically not very scientific, that is, there is usually no database available to actually calculate probability from a statistical perspective. Much of risk assessment is based on prior experience and a full detailing of project tasks and components so that they can be understood.

TABLE 4.3 Risk Matrix Template

Risk item	Description of risk	Impact (technical, schedule, cost, quality)	Severity (high, medium, low)	Rank	Contingency plan (what is planned to offset the risk)

2. *Sensitivity.* Sensitivity analysis involves analyzing how sensitive the project is to a particular risk, that is, quantifying linkages between a given risk and the "ripple" effect it could have on the project from a schedule, quality, cost, and overall performance perspective.
3. *Decision tree.* A decision tree helps to describe the flow of project decisions to points where there are tradeoffs based on risk. That is, it clarifies where you have to make decisions based on a risk and other factors, such as cost and quality. For instance, an important leg of a decision tree on planning a project task may be to "buy or make" a particular component, that is, to contract-out or to build in-house. The risk implications are different for each. Thus risk is a part of making that decision.

Risk matrix template

The risk matrix is a very valuable but simple tool to manage the risk "portfolio." Table 4.3 shows a risk matrix template. Chapter 6 contains more examples.

Build a Risk Management and Planning Process

Building a risk management process is largely a "selling" challenge, not a technical task. Building support systems for risk management is part of the process by which the organization matures as an organization.

Given a project description and ranked list of risks, describe the steps in the project risk management process and develop a risk response strategy including avoidance, transference, mitigation, and acceptance, and develop a risk response plan.

This has to do with responding to the risks we have identified with a process and risk response plan. This is where the project manager produces a strategy—a way of addressing the risk identified. For instance, for the software debug process referred to above, I must develop a plan to avoid the risk and/or respond, and it might include alerting the software developer to the risk and getting a direct estimate of his or her view of impacts, as well as an approach to offsetting the risk—perhaps assigning another developer to the job or putting more effort into debug or even subcontracting out a support function to anticipate debug problems during in-house design.

This covers front-end, company-wide risk management system planning as well; you not only have to describe a risk management system, but you need to install it into the organization as a way of doing business. The assumption goes like this: If your company or agency is going to identify and manage risks, it must have the capacity to use the right systems, tools, and formats to do so and it must have a well-defined process as well. So you have to establish a risk planning capacity and process. This new process was added in the most recent version of the PMI's *PMBOK* to address "getting your company ready" to do good risk management. In sum, we address organizational capacity as the ability of the company or agency to develop a company-wide risk management process to help project managers mitigate (offset) project risk and develop and implement a risk response and communication plan.

Risk management plan

The risk plan does not always need to be documented; it is the thought process that counts.

The risk management plan will include the following sections, which will be submitted to the project manager.

Risk management policies and procedures define the way the company or agency intends all its project managers to approach risk. The company or agency in effect says to the project manager, "here is the way we want to ensure consistent approaches to risk management across the company, and if you stay with these guidelines, we will support you even if the project does not produce exactly as planned."

Here we address the project risk management process. The basic steps in the project risk management process, which are mirrored in the structure of our course, are:

- *Risk planning:* the step that *prepares* the organization for risk with support systems and defined roles and procedures, setting a high priority on risk management as an integral part of the project management process
- *Risk identification:* the step that identifies risk and ensures that all risks are "on the table" before more detailed assessment is undertaken
- *Risk assessment:* the step that assesses risks, analyzes them, and quantifies risks in terms of probability
- *Risk scheduling:* the practical application of MS Project, PERT analysis.
- *Risk response:* includes the development of a risk management plan, contingency plans, and reflects risk in the project scheduling and resource assignment process
- *Risk monitoring and communication:* the step that provides for regular reviews and interchange among key stakeholders and customers on risk, changes in risk, and risk-related developments

Here is where the company defines the roles of project managers and team members, such as project engineers or software developers, in defining and controlling risk. Since risk assessment is an analytic process and since risks change over time, the management of risk takes time and effort. So jobs that involve risk management must be defined as such, and team members must be enabled to charge time to risk management.

Work breakdown structure. Remember, a good risk management plan starts with a well-defined project deliverable accomplished through a WBS. The WBS presents the outline of the project deliverable. Thus, if it is comprehensive and does not omit major components and work packages, it lists the potential sources of risk.

Risk identification

Here is where we address the risk identification process. It starts with a full understanding of the project and a comprehensive project plan. Risks are identified from the work breakdown structure and baseline schedule. Risks are categorized in terms of technical, technical and equipment processes, quality, organizational, people resources, project management processes, or performance risks.

Tools for risk identification include brainstorming (group thinking out loud), Delphi technique (bringing experts together), interviewing, checklists, assumptions, and process diagramming techniques. They may also include drawing on lessons learned from similar projects in the organization.

We want the reader to get: (1) a full understanding of how a company or agency equips itself to handle risk in its projects—learning what management must do to build the capacity of the whole organization to control risks and (2) a full understanding of how individual project managers handle the initial step of identifying risks since this first step is so critical. If you miss major risks upfront, handling them downstream in the project becomes more costly and sometimes just doesn't work.

Our objective, then, is to learn how to categorize risks and we will use a risk matrix template to do so. A risk matrix template is a table showing categories of risk down the left column and providing information on each risk across the top. One approach to this information is to list definition, potential impact, probability of impact, response plan, and monitoring plan. Another approach categorizes risks in terms of external unpredictable (like our earlier definition of uncertainty), external predictable, internal nontechnical (I would call this managerial/organizational), technical, and legal.

We will go deeper into risk identification, identifying and categorizing risks in various kinds of projects. This is where you will see "windows and flags" for identifying risks and getting them "on the screen" early in the project. For instance, in a highway construction project, a rule might be that you can always predict that subcontractor concrete finishers represent a major risk (reliability, quality of work) and need to be listed in the risk matrix. If you know you have a major challenge up front—not because it's a particularly difficult task but because its history suggests uncertainty—then use risk management tools to define it and monitor it. This is a risk "flag" in that it flags a potential risk simply by its definition. It's also a "window" of opportunity to get on top of that task early.

The success of a risk identification process lies in the initial project work breakdown structure. If you have identified all the necessary work to compete the deliverable in the WBS then risks can be identified by looking at each risk and going through the mental exercise of asking, "what could go wrong and what would happen if it did?"

The real problem results from not identifying a task in the WBS that later creates real risk exposure—missing a major risk in the initial stages because you missed an important function or component. If it isn't visible early, a major risk can "show up" later. For instance, you may have identified a major software engineering task in your WBS that isn't that challenging, so you assign a low risk to it. But the real risk lies in the availability of a key software engineer to do the work—which you did not identify in the WBS and which later gives you real headaches!

Communicate risks

Communicating and reporting on risks is part of communicating and reporting on the project itself. You are not only trying to reduce uncertainty and report accomplishments, but also to alert stakeholders that the project is in a sensitive stage of uncertainty.

Given a list of risks, you will need to know how to develop a plan to communicate those risks to the appropriate project team members and stakeholders, as well as to management.

This reminds us that it doesn't make much difference what risks you have identified if you haven't communicated with people who have a stake in the project and who can help address risks. This requires us to identify those people who have a direct interest in the project outcome and who must commit to the risk management process. These interested parties would include the customer or client, all project team members, support personnel, corporate management, investors, and even the competition.

The challenge here is to figure out how to present a risk so that people understand it and come up with good ideas on how to handle it. You don't want to create fear; you want to create commitment and resolve. You also want to focus key people on the right risks, the ones that can create the most problem and opportunity. The 80 to 20 rule says that 80 percent of your risks will not be important; it's the 20 percent that will be critical that you need to identify and manage.

Monitor and review risks

Keeping up with risk involves equipping your team members and suppliers with the tools necessary to monitor risk and the sensitivity to catch risk problems before they occur. It is mostly an interpersonal process, not a formal project review process.

Given a project in which tasks have been scheduled and risks have been identified and ranked, develop a plan to monitor and control changes in risks and risk potential as an integral part of the project review process.

It takes some planning and effort to monitor how risks change as the project progresses and how those changes affect the original risk assessments and project outcomes. This is typically done through a project review process where individual project risks are reviewed as part of the broader review of the whole project. For instance, in our software debug issue, an upcoming software design review might be scheduled and during the project review prior to that design review a checklist might be constructed to "flag" debug issues.

The other part of the monitoring process is making sure that the team members closest to the job are aware of risks and communicate risk information as the design and development is occurring. They are the most likely to know the extent of change in a risk assessment and how it will impact the project.

Develop risk-based schedules

The risk-based schedule referred to earlier is simply a schedule, which is "informed" by risk considerations. Durations reflect things that could happen.

Given project risks, you will need to know how to develop a PERT chart for the project showing worst case, best-case, and most likely case schedules, based on risk.

This subject is a special one, focusing on building your skills to use the PERT chart tools in Microsoft Project or equivalent project management software—to identify and weigh worst, best, and most likely cases, and to write notes that will assist in project reviews and communicate risks through the software.

We are not expecting you to be experts in the PERT charts, just to know how they are used. If you are going to be a project manager dealing with risks in a project, you will have to know the basics of project management scheduling and PERT analysis.

Risk scheduling. Project schedules serve several purposes:

1. Schedules provide a transition from project definition, such as the WBS, to time and cost, and to interdependency. That is, schedules pin down actual tasks, durations, start and finish dates, and resource assignments, and most importantly, linkages between tasks. You manage from a schedule, not a WBS.
2. Schedules enable a project manager to lay out the total project impact of a given risk scenario, that is, given a number of complex tasks and interdependencies (links), creating different durations and resource mixes for selected tasks which involve risk gives the project manager a way to measure schedule impacts.

This is why we have you identify three scenarios using the PERT tool in Microsoft Project, pessimistic, optimistic, and expected, for the five greatest risks in your project. This process helps you see the usefulness of having three schedules with different milestone dates and seeing that when you assume an optimistic scenario you may have scheduled a given resource to do the work, but in the pessimistic scenario that resource is no longer available! Now you find that the worst case impact of one risk has created other risks because it has pushed the schedule into a window where the original schedule assumptions will not work.

A Note on Microsoft Project PERT and Risk Matrix Terminology

Microsoft Project uses the terms *pessimistic, expected,* and *optimistic*. Expected usually means the duration you originally estimated without concern for risk, although it may not. Optimistic means the risk and impact are low and you think you might be able to "beat" the expected. Pessimistic means that the risk and impacts are high and you don't think you will be able to "make" the expected duration.

In the risk matrix you use the terms *high, medium,* and *low* for risk rankings. In general, high is over 50 percent probability and high severity; medium is less than 50 percent probability and moderate severity; and low is less than 10 percent probability and low impact.

Table 4.4 compares the terms from your risk matrix and your PERT analysis.

Risk Response

We will focus on risk response, monitoring, and communication. Here is where we get to the "meat" of the project risk management process—the planned and scheduled action that a project manager takes to deal with project risks, the review and monitoring of changes in risks and risk management effectiveness, and the communication and interchange between key project participants on risk issues.

By now, you should have a good grasp of what project risk assessment is all about. "So what?" you say. Well, now we will answer the question, "Once I know what my risks are going to be and what impacts they may have, what do I do about it?"

There are four parts of the response process— response, monitor, review, and communicate.

1. *Response.* Response involves developing a risk management plan to address the risks identified and rank ordered in the project plan. Response means control,

TABLE 4.4 Comparison of Risk Matrix and PERT Analysis

Risk ranking in risk matrix	High (risk severity is high and the probability that it will happen is high)	Medium (risk is moderate and impact not so severe)	Low (risk is low and impact low even if it occurs)
Microsoft Project terminology	Pessimistic (duration reflects concern based on the probability that risk will occur and will have major adverse impacts and slip the schedule)	Expected (original estimate of duration without considering risk, unless there is a reason to change it)	Optimistic (duration reflects low risk and therefore "hope" that the risks can be controlled by contingency plans and the task can be completed quicker than expected)

e.g., assuring that there is a contingency plan for each risk that might be triggered by a change in a key indicator, such as meeting an intermediate milestone in the design of a piece of equipment. This is the key part of project risk management—the development of a risk management plan that provides you a prepared response to anticipated risks, tailored to that risk and its impacts. This is why we have you develop a risk management plan as part of the course.

2. *Monitor*. Monitoring involves identifying indicators that "flag" that a given risk may be occurring and that the probability of a given risk has changed. For instance, if the risk involves a technology performance issue (a given piece of equipment must perform to a standard) then the earliest indicator that the equipment is not performing should be identified through a monitoring process. In this case, an engineer or contractor is requested to submit a report on a preliminary "test run" of the equipment, early enough to head off a major disaster when the equipment fails in use.
3. *Review*. The project review process involves establishing an agenda and process for regular reviews. The agenda reflects risks and requires that each review meeting specifically address the risks inherent in a project.
4. *Communicate*. Communicating risks to the right parties improves the capacity to respond to risks and also offsets the tendency to have an isolated view of risk within the team. Communication involves reporting the results of risk assessment, the risk management plan, and monitoring and review results to key stakeholders including:
   a. Customer
   b. Investors
   c. Management
   d. Project team
   e. Contractors

Risk response, monitoring, and communication are integral to the process, not a separate activity. In the real world of project management, the response to risk is played out in a series of day-to-day decisions as developments unfold, guided by the risk management plan. Risk response is really a state of readiness, a focus that the project manager has on certain aspects of the project. The focus changes over the course of the project because key risks come in and out of focus. The purpose of risk management planning is to narrow down the critical "things" to keep your eye on and manage during the life cycle of the project.

Risk response factors

The capacity to respond to risk is heavily dependent on the quality of planning and control in the project cycle. Risks that have not been anticipated and contingencies that have not been integrated into the schedule will be difficult to respond to, regardless of intent. This means that risk response must be timely and embedded into the planning process, beginning with the WBS and schedule. There are several factors that weigh heavily on the success of risk response (Fig. 4.1).

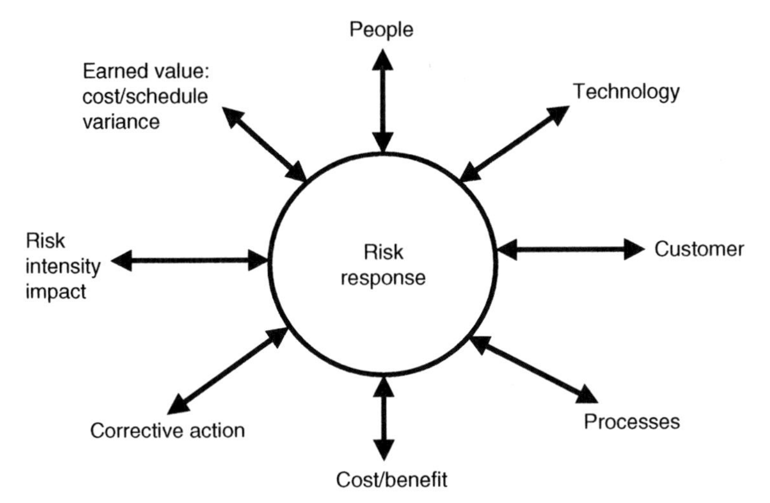

Figure 4.1 Risk response factors.

Earned value. Cost and schedule variance are key inputs to risk response, particularly root causes. The root causes of schedule slippage and cost escalation are typically related to underlying risks that could and should have been addressed in the up-front planning process starting with the WBS. Risk planning should uncover potential root causes before the final baseline schedule is completed. Earned value can be seen as an indicator that a risk has not been attended to; a late wake-up call that is often difficult to address.

Risk intensity. Despite quantitative tools for calculating probabilities and intensities, there is considerable personal and professional judgment involved in aligning a risk response with the intensity and impact of that risk. In the end, a project manager must make the decision on the expected value of a given decision at a key project milestone or crossroad. Intensity is best dimensioned by those closest to the point of impact (e.g., customers, project team members, functional and technical professionals), so the most accurate source for evaluating intensity is to tap the judgment of the key stakeholders.

People. As we have pointed out earlier, risk management is essentially a people issue, not a technical or analytic issue. Project managers must depend on the awareness and judgment of those closest to the work to assess and respond to risk. The key is placing responsibility for risk identification and response in the hands of those planning and designing the work early enough to incorporate all worst case scenarios. Key stakeholders will respond to a culture that encourages and enables early contingency planning by anticipating risks and integrating them into project schedules and budgets. In the absence of such a culture, the organization can quickly deteriorate into finger pointing when unanticipated risks impact a project schedule or budget.

Technology. Technology creates risk simply because projects often involve designing and testing new technologies or integrating new systems. New systems and products are by definition risky and involve key decision points based on risk. But these risks should be addressed in the design and testing process itself. In other words, technology risk is integrated into every step of the design process. For instance, in the development of a new electronic instrument, the project team faces the risk that the instrument will fail in tough industrial applications. That risk is translated into a design function to test the instrument in those conditions, thus responding to the risk at the point of design.

Customer. The customer is a major factor in responding to risk since it is the customer who will likely foot the bill for response and pay the price for unanticipated risk impacts. Thus the customer must be involved in every step, from concept through prototyping and production. The most practical approach here is to report to the customer on the risks in a project specifically in terms of customer expectations. That involves understanding customer expectations and reporting on progress against those expectations regularly.

Processes. Often, a key process will determine how a risk is addressed and how risk response is managed. For instance, if prototyping is built into a system's development process, the risk of customer rejection of the product can be offset early. It is in defining the process that key risks and responses are *designed into* the way work is done.

Cost/Benefit. Two kinds of costs occur in risk management—the costs of responding to unanticipated impacts of risk and the costs of planning and addressing risk proactively as part of the process. It is not the cost of risk response that determines next steps as much as the *relationship* between costs and benefits, and the *timing* of the response. "Pay me now or pay me later," might well be the guiding principle. The cost of a given product test in the design process is likely to be much lower than the cost of responding to a product failure on a performance standard that was not tested in design.

Corrective action. Despite the best-laid plans and schedules, project management is largely a process of midstream adjustments and corrections based on the actual dynamics of a project in progress. That means that close monitoring of key project indicators is key to real-time response to actual progress. Corrective action, the process of continually bringing a project into line with its performance objectives, and aiming the process to completion, is facilitated by contingency plans already built into the schedule in anticipation of risks and uncertainties. Without such plans, corrective actions are often ill conceived in the heat of the crisis and often generate counterintuitive results. What is expected as a result of a given action not only does not happen, but other things happen in the project as a result, which create new problems.

Contract Management

Various kinds of contracts can have substantial impacts on the success of risk management strategies. As shown in Fig. 4.2, the buyer and seller risk is associated with a variety of contract types. Generally, fixed price contracts create risk for the seller, while cost reimbursement contracts create risks for the buyer. Time and materials (T&M) contracts fall somewhere in between.

Given project risks and the need to contract project work that is at risk, develop a strategy for using appropriate contract types to reduce and/or share risks with project contractors.

Some of your biggest risks may be encountered in managing contractors who are doing some of your project work. Thus it's important to choose the right kind of contract type and process to make sure the contractor shares in the risks of completing the work. The more you can get the contractor to share risks, the more likely that the contractor will do a good job and flag risks early. We will explore fixed price, unit price, and cost reimbursable contract types and their impacts on risk management.

In sum, the book takes you through a process, beginning with definitions, early identification and categorizing of risks in your project, moving to assessing those risks, and then building risk planning capacity in the organization. Then we cover the process of communicating and reporting on risks to various stakeholders, and then we cover monitoring and reviewing risks. Next we address scheduling risks and working on pessimistic, optimistic, and expected scenarios and then we address how we respond to risk, prepare risk management plans and communicate them. Simply put, the course follows the risk management process—risk identification, risk assessment, risk planning, risk reporting, risk scheduling, and risk response.

When you think about it, the decision to contract out work changes the risk "equation" since work done in-house does not have the same "leverage" that

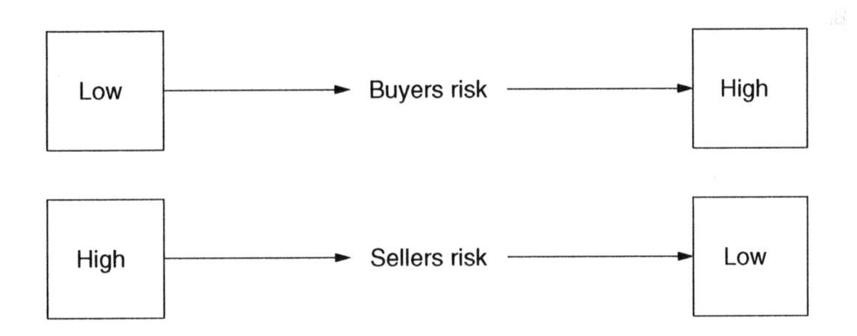

Figure 4.2 Risk and types of contracts.

contracting out provides. First, you can rarely "incentivize" (provide incentives, such as added compensation) in-house personnel to manage risks successfully on a particular task, while a contract allows you to do so. Second, contracts allow you to document shared risks, that is, to share with the contractor the costs of a particular risk occurring and impacting the project.

Various kinds of contracts have different impacts on contractor performance because the nature of the contract sets the conditions for the work. Contract type can impact financial objectives, contractor involvement, costs, schedule, and quality, as well as customer satisfaction. Here are the four basic contract types and their implications for risk.

Lump Sum Contracts. Lump sum contracts encourage the contractor to cut corners since there is no process for getting more funding. The risk is that if the deliverable is not well defined, the contractor may not produce to the project design. Lump sum puts all the risk on the contractor, and therefore the contractor acts accordingly.

Unit Price Contracts. Unit price is based on paying for each unit. This kind of contract shares risk with the contractor, encouraging the contractor to reduce the cost of volume production.

Target Cost Contracts. Target cost is the ultimate shared-risk contract, since both parties estimate a cost and work together to achieve it, leaving the door open for more funding if the target is not achieved.

Cost Reimbursable Contracts. Cost reimbursement contracts do not share risks very well. The contractor can repeatedly claim more costs based on continued work to refine the deliverable and cover past mistakes. The project manager is left with the basic risk in a cost-type contract.

The process of negotiating a contract brings out the risk issues because it is in the interest of both parties to avoid surprises and anticipate and mitigate risk. Financial and costing considerations become the vehicle for sharing risk. In the process, the contractor attempts to minimize risks by pressing for more clarity in requirements and definition of the deliverable and by negotiating a contract type and price that protects the contractor from unforeseen risks. The contracting company or agency attempts to minimize risks by transferring them to the extent feasible to the contractor. The contract relationship helps to clarify the implications of risk and how risks can be avoided.

Chapter

5

Making Risk Policy: A Risk-Based Program Management Manual

The purpose of this chapter is to provide an example of a manual that might be useful in setting the stage for effective risk management.

The purpose of the manual is to describe a risk-based program management system. The term *program management* describes the planning, scheduling, tracking, and delivery of programs and projects through the product development process. A *program* is a product line or portfolio and a *project* is a phase of a program. For a product development process, the process typically produces:

1. A product documentation package
2. A product prototype and documentation for a product demonstration
3. An engineering design

Principles embody policy and guidelines for the program management process and risk is to be handled. For instance, it would be policy that all programs and projects will be planned and managed in a way consistent with these principles. The process is managed by a program manager who is responsible for producing products that satisfy customer requirements. Customer requirements are documented in a *system requirements specification* (SRS). Products are produced through a product development process, which meets customer requirements. In collaboration with department managers, program managers follow this manual in planning, scheduling, and tracking programs through the product development cycle and in identifying, assessing, and responding to risk.

The following principles underlie the risk-based, program management process.

Meet customer requirements

A risk-based planning process is customer-driven, striving to meet or exceed customer requirements and control customer risk. Typically, the customer's technical requirements are embodied in an SRS prepared by the system engineer. The customer's schedule and resource requirements are embodied in a baselined, program schedule prepared by the program manager and approved by the director of product development. In addition, the program manager maintains close contact with the key product manager or contact point for a program to ensure that customer needs, expectations, risks, and scope changes are addressed and managed through a systematic process.

Follow product development process

Risk is in the details, in the technical process of producing a product or service. The program management process will ensure that products are managed through the product development process tailored to particular product requirements and risks. Program managers use the work breakdown structure (WBS) in that policy as the basis for scheduling a program with exceptions for special programs, which do not involve the entire certification WBS.

Standard work breakdown structure

The standard WBS for schedules is specified as follows:

- *Program.* This is the product line incorporating a basic set of features and functionality.
- *Project phase.* This is the particular set of features and functionality for a program, based on particular customer needs and risk tolerance (level of capacity to absorb risk).
- *Stages.* These are the generic steps in the product development process, e.g., requirements, detailed design, prototype development, design validation, verification, and manufacturing transition, all containing risks and decision options.
- *Functions.* This is the task level within a stage, e.g., mechanical, electrical, and software design within detailed design.
- *Tasks.* This is the operating component level where work is achieved through individual or small team activity and where risk is built into the work of people who do the real work.

Teamwork

Program managers and functional managers establish teams to carry out the work with the objective of building an environment of high-performance teamwork and collaboration. Teams address risk as part of their everyday work.

To the extent possible and to avoid risk inherent in bad performance, staff will be assigned tasks that are consistent with their backgrounds and expressed professional interests. Program teams will be composed of professionals who are suited to the work they are expected to perform. Team staff members will be oriented and trained as necessary to enable them to perform assigned tasks, again to avoid risk.

Define and communicate the scope of work, risks, and assignments clearly

Product requirements and job assignments will be defined and communicated to the program team clearly through the program schedule and individual assignments. This is to empower staff to understand how their work contributes to the overall customer requirement and to prepare for work assignments with appropriate training and development.

Collaboration across the organization

Collaboration between the program managers, department managers, and the program team is the essential ingredient to the success of program management. The organization encourages continuous, professional communication and information exchange among program team members and department and system managers in a concurrent engineering framework. The objective is to create both individual team member accountability for particular tasks and inherent risks and a broad support system to ensure individual success despite risk.

Work will be risk, quality, cost, and schedule driven

Maximum emphasis will be placed on preparing tight program schedules that incorporate all the work necessary to meet requirements on time. Program schedules will be planned and "scrubbed" in a collaborative process that ensures that all necessary work is included and all task durations, interdependencies, and resources are tightly planned and estimated. Schedule baselines will be established and work initiated only after schedules have been tightened through this process.

Ensure timely procurement of product components

Program managers pay special attention in early program scheduling to ensure the availability of required product components and test equipment. Hardware specifications, parts, test equipment, and supply items will be included in initial program schedules and appropriate lead times established. Risk planning and control are integrated into the schedule as well as contingency plans. Procurement actions will be generated in a timely way to avoid schedule delays attributable to lack of components.

Change will be managed

The organization will administer the engineering change notice and configuration management processes to ensure that requirements and product component changes are managed and controlled. A systematic change management process ensures that specifications can be met within schedule and resource constraints.

Progress in reducing risk will be tracked periodically reviewed

Program managers will track program progress, prepare weekly reports, and prepare for weekly program reviews conducted by the director of product development. Department managers and selected team members will participate in risk tracking and program reviews as appropriate.

Program Management: Roles and Responsibilities

The organizational framework for program management is a collaborative, team-based organization requiring cooperation between program managers and functional department managers. In that framework, the program manager has responsibility for delivery of the product within quality, schedule, and resource requirements. Department management is responsible for supervising staff and maintaining the technical capacity of their departments to support program management. There is a general awareness of risk issues and daily discussion of potential for overcoming risk.

Program management office

The program management department includes all program managers and the program administrator/planner. The department is responsible for assuring consistency in the application of this program management guide throughout the product development process. The department tracks performance of the overall program management process.

Within the department, program managers will consult regularly with each other on scheduling and resource plans and potential conflicts, and ensure consistent approaches to scheduling details, work breakdown structure, budgeting, and sharing schedule information. On the initiation of new programs, program managers will consult with each other on potential resource impacts and issues.

Program manager role

The program manager is ultimately responsible for meeting customer requirements, managing risk, and delivering the program within schedule. The program manager provides leadership to the program team, ensures that the program meets product specifications, delivers the product on time and within resource constraints, and in general controls the "what and when" of the project. The program manager produces time-phased schedules for each program, tracks progress and anticipates future impact, and ensures linkages with related programs and projects.

The program manager manages programs through all stages of product development and, along with the systems engineer, ensures that required design reviews are conducted and documented, and all actions resolved.

The program manager has the primary responsibility for creating a program plan for each program and a program schedule composed of tasks and milestones. The program plan is created with support from the customer, program team members, and department managers. Once the program is underway, the program manager is required to keep the program schedule current, track progress, and incorporate changes as required. The program manager uses project management software to produce and update schedules and resource reports, and is expected to be proficient in the use of such software for control and presentation purposes.

The program manager has the primary responsibility of creating and maintaining a detailed program schedule that meets all program objectives. The schedule must be consistent with the generic WBS, and includes:

- Summary tasks and task structure and key milestones that correspond to all major program objectives contained in the program plan.
- All product development activities and tasks required to execute a given program including systems design, detailed design, certification, test equipment, reliability, safety, design reviews, manufacturing, procurement, and test assets.
- Tasks detailed to the lowest practical level. Activities and tasks should generally be built four levels down.
- Resources assigned to activities and tasks and leveled to reflect a realistic workload.

Departmental manager roles

Department managers for systems engineering, mechanical/optical design, electrical design, software design, and certification are responsible for building and maintaining the resource and technical capacity of their departments to support the product development process. Here are some key functions of department managers in the program management process:

- Assign staff to programs and support assignments
- Ensure technical processes and systems are in place to identify and address risk and accomplish programs
- Prepare and maintain department schedules for each program, identifying department level assignments
- Attend program review meetings and provide advice and support to program managers

Program team member roles

Risk management starts with people and how they approach their jobs. Each program team member is responsible for understanding his or her individual tasks and inherent risks, and for general support to overall team performance. Team members are accountable for keeping technically proficient and performing their assigned tasks in a timely way, consistent with the schedule. Team members are responsible for communicating with their program managers and functional managers on issues or problems encountered in their team tasks. They collaborate with each other and the program manager, promptly attend program team meetings, and report to their program managers and department managers on schedule and technical issues, respectively.

Role of the program administrator/planner in the PMO

The role of the program planner in the program management office (PMO) is to promote consistent best practice in program and risk management. The administrator/planner provides administrative support to program managers and departments with scheduling, resource planning, and reporting service, and prepares analyses of resource impacts to identify and resolve conflicts. In addition, the program administrator/planner prepares program management guidelines, provides training, and develops program evaluation metrics, and maintains individual program schedules for the director of product development and/or program managers.

Program Planning, Scheduling, and Resource Management

The product development process is primarily schedule-driven. Effective and disciplined scheduling and tracking of work and resources is directly related to customer satisfaction since customer expectations always include timely delivery as a key priority.

The scheduling process begins with a program plan that describes the overall program in general terms. It is a reference source for all documents which impact on the program. The program plan includes:

- Program overview
- Program strategy
- Customer identification
- Program objectives
- Risk management
- Measures of program success
- Program scope and requirements (Summary)
- Program management, including team roles, schedule, resource plan, and milestones, program review, and risk management
- Program development and review process
- Reference documents

Once a program plan is approved, good scheduling is at the heart of the program management process. Program schedules created in Microsoft Project, linked to a central resource pool file, and posted on the network constitute the basis for program development, tracking, and review. A good scheduling process provides adequate time to ensure that the work breakdown is comprehensive and responds to the customer requirement and product functionality, that scheduled task durations and predecessors are as accurate as possible, that key linkages are made, and that people and resources, once assigned, understand interdependencies and are available and committed to the program when they are needed.

Before a schedule is drawn up, the work itself must be clearly defined in a WBS. Thus scheduling provides for an SRS (in the *requirements stage*), which defines the *what and why* of the program or product (e.g., a description of the product and its functionality and its inherent risks). The scheduling process helps to flesh out requirements as individual features are programmed into various iterations of the schedule, leading to the baseline.

Once the work scope is understood and signed off by the customer, scheduling defines *when and how* the work is going to be done, key interdependencies, *when* the deliverable will be produced, and *who* will do the work. The scheduling process assigns staff to scheduled work and commits staff to do the work within the time constraints in the schedule. Scheduling is a resource planning tool providing a high degree of discipline to the assignment of staff since each task is specifically described and time-constrained. Thus scheduling requires that those who actually going to do the work—*those who are being scheduled*—also be part of the process. Since scheduling "signs up" and mobilizes staff to fit new work into their schedules, which typically include other program work, it requires that there be a clear picture of staff availability (e.g., the current resource picture). New program schedules are phased-in based on the timelines and resource impacts of current work.

The integrity of a schedule is only as good as the description of the work, the processes in place to do the work, a good picture of interdependencies and resource availability, and the commitment of the people who are slated to do the work. This process becomes more complicated when there are several programs or projects operating at the same time—a multiproject environment—where staff time is always limited by previous commitments driven by earlier or concurrent projects, and by anticipated and unplanned work. In the end it is the quality of the planning and communication process and the capacity, commitment, and motivation of the individuals actually doing the work that drives a successful schedule.

Project management software makes it easier to accomplish the scheduling process by capturing important planning and scheduling data and making it available to a wide cross section of people, and facilitating presentations and progress tracking. The following process assumes access and proficiency in Microsoft Project as the support software as a network-based planning and communication tool.

Scheduling tailors the product development process to real time and available resources, and flags conflicts and new resource needs. Scheduling provides time-phased and linked tasks and milestones, assigns resources to complete the work, and supports the monitoring of performance, resource allocations, budget, and earned value.

Figure 5.1 is an example of a baselined program schedule (Gantt chart). Note the inserted column for *percent complete*, added through the insert toolbar, then column. Note that predecessor for ID 88 is 87FF+1, entered through the task information dialogue box. 87FF+1 indicates that the task has a finish-to-finish relationship with ID 87, plus one day.

Figure 5.2 is the resource usage view of that same schedule. The program manager and the project team can see from this view what resources are assigned to the project and the level of assignment in terms of hours, against a calendar.

Here are indications of risk bottlenecks in how resources are shared between projects. Applying the theory of constraints and critical chain management, program managers see project bottlenecks early and focus on managing scarce resources, and use project buffers, to control key people or equipment that could later delay project delivery.

Figure 5.3 is the tracking Gantt view of the schedule after it has been updated. Note that the tracking view includes a percent complete column with actual completion percentages.

ID	% Cor	❶	Task name	Duration	Start	Finish	Predec
82	0%		453210-950 Chassis Detailed Design	20 days	Tue 4/24/01	Mon 5/21/01	81
83	**0%**		**RCU Optics Assemblies**	**93 days**	**Mon 1/15/01**	**Wed 5/23/01**	
84	0%		RCU Optics Preliminary Layout	40 days	Mon 1/15/01	Fri 3/9/01	
85	**0%**		**453110-950XXX Bezel Assy**	**16 days**	**Tue 3/13/01**	**Tue 4/3/01**	
86	0%		453110-950 Bezel Assy Design	10 days	Tue 3/13/01	Mon 3/26/01	70
87	0%		**Send Bezel Assy to Tucson/RDE**	0 days	Mon 3/26/01	Mon 3/26/01	86
88	0%		**Receive Approval from Tucson/RD**	0 days	Mon 4/2/01	Mon 4/2/01	87FF+1
89	0%		**453110-950 Bezel Assy Release**	1 day	Tue 4/3/01	Tue 4/3/01	88
90	**0%**		**RCU LCD Assembly**	**30 days**	**Wed 4/4/01**	**Tue 5/15/01**	
91	0%		453122-950 RCU Front Filter	3 days	Wed 4/4/01	Fri 4/6/01	85
92	0%		453115-950 Tray Design - Prelimin	5 days	Mon 4/9/01	Fri 4/13/01	91
93	0%		453180-950 DIB Outline	5 days	Mon 4/16/01	Fri 4/20/01	92
94	0%		453127-950 Bezel Flex Outline	5 days	Mon 4/16/01	Fri 4/20/01	92
95	0%		453120-950 Assembly Drawing	3 days	Mon 4/23/01	Wed 4/25/01	94,93
96	0%		453129-950 LCD Altered Item Draw	1 day	Thu 4/26/01	Thu 4/26/01	95
97	0%		453121-950 LCD Mount Adhesive	3 days	Fri 4/27/01	Tue 5/1/01	96
98	0%		453128-950 Diffuser Design	3 days	Wed 5/2/01	Fri 5/4/01	97
99	0%		453115-950 Tray Final Design	2 days	Mon 5/14/01	Tue 5/15/01	**123**
100	**0%**		**RCU Backlight Assembly**	**13 days**	**Mon 5/7/01**	**Wed 5/23/01**	
101	0%		453142-950 Lamp Design	5 days	Mon 5/7/01	Fri 5/11/01	98
102	0%		453132-950 Front Element Design	3 days	Mon 5/14/01	Wed 5/16/01	101
103	0%		453136-950 Lamp Housing	5 days	Thu 5/17/01	Wed 5/23/01	102
104	**0%**		**RCU Assembly Drawings**	**50 days**	**Tue 4/24/01**	**Tue 7/3/01**	
105	0%		453290-950 Rear Panel Assembly Draw	2 days	Tue 4/24/01	Wed 4/25/01	78
106	0%		453270-950 I/O #1 Assembly Drawing	1 day	Tue 7/3/01	Tue 7/3/01	128
107	0%		453250-950 LVPS Assembly Drawing	1 day	Tue 5/29/01	Tue 5/29/01	132

Figure 5.1 Baselined program schedule.

ID		Resource name	Initials	Details			March					April
					2/11	2/18	2/25	3/4	3/11	3/18	3/25	4/1
11		Joe Allen	SA	Work	40 h	40 h	40 h	40 h	32 h	40 h	8 h	32 h
		RCU Optics Preliminary Layo	SA	Work	40 h	40 h	40 h	40 h				
		453110-950 Bazel Assy Des	SA	Work					32 h	40 h	8 h	
		Sand Bazel Assy to Tucson	SA	Work							0 h	
		Receive Approval from Tucson	SA	Work								0 h
		453110-950 Bazel Assy Rele	SA	Work								8 h
		453122-950 RCU Front Filter	SA	Work								24 h
		453115-950 Tray Design - Pr	SA	Work								
		453180-950 DIB Outline	SA	Work								
		453127-950 Bazel Flex Outline	SA	Work								
		453120-950 Assembly Drawing	SA	Work								
		453129-950 LCD Altered Item	SA	Work								
		453121-950 LCD Mount Adh	SA	Work								
		453128-950 Diffuser Design	SA	Work								
		453142-950 Lamp Design	SA	Work								
		453132-950 Front Element	SA	Work								
		453136-950 Lamp Housing	SA	Work								
12		Bill Wolff	BS	Work	40 h	40 h	40 h	40 h	32 h	40 h	40 h	40 h
		453210-950 Chassis Pralimin	BS	Work	40 h	40 h	40 h	40 h				
		453270-950 I/O #1 CCA Out	BS	Work					16 h	4 h		
		453250-950 LVPS CCA Out	BS	Work					16 h	4 h		
		453230-950 CPU CCA Outline	BS	Work						16 h	4 h	
		453280-950 I/O #2 CCA Out	BS	Work						16 h	4 h	
		453225-950 Ballast Drawing	BS	Work							32 h	8 h
		453211-950 Rear Card Guide	BS	Work								16 h
		453291-950 Rear Panel Des	BS	Work								16 h
		453210-950 Chassis Detaile	BS	Work								
13	◆	**Dennis Hsu**	**DH**	Work								

Figure 5.2 The resource usage view.

Figure 5.4 is the tracking Gantt with bar chart, which shows actual progress (black bar) within the planned duration (blue bar).

Five-step scheduling process

The program manager generates the scheduling process, the department manager serves as a resource on product functionality and department resources and ensures that the technical procedures are in place to complete the work. This process works effectively only with a constant dialogue between program managers, department managers, system engineers, and the team.

The scheduling process for product development involves five steps culminating in the product deliverable. The general sequence of work is first to define the work from customer requirements; structure the work into an outline or work breakdown structure; define an overall, top-level task structure and work flow; then identify tasks, durations, risks, and interdependencies, and then develop

ID		Task name	Duration	Start	Finish	Prede	% Complete
1		**Stage 1 Systems design and requirements definition**	**195 days**	**Fri 3/3/00**	**Fri 12/1/00**		**33%**
2		**Program management**	**136 days**	**Fri 3/3/00**	**Mon 9/11/00**		**95%**
3	✓	Detailed program schedule	6 days	Fri 3/3/00	Fri 3/10/00		100%
4	✓	Manpower planning	6 days	Fri 3/3/00	Fri 3/10/00		100%
5	✓	Schedule management review	4 days	Tue 6/13/00	Fri 6/16/00		100%
6	✓	Program plan (Internal)	2 days	Mon 6/19/00	Tue 6/20/00		100%
7		Schedule baseline established	1 day	Mon 9/11/00	Mon 9/11/00	6,5	0%
8		**System engineering**	**56 days**	**Mon 8/28/00**	**Tue 11/14/00**		**60%**
24		**Reliability engineering**	**10 days**	**Mon 9/11/00**	**Fri 9/22/00**		**0%**
27		**Safety engineering**	**50 days**	**Mon 9/25/00**	**Fri 12/1/00**		**0%**
32		**Stage 2 Detailed design**	**302 days**	**Fri 11/19/99**	**Tue 1/16/01**		**44%**
33		**Electronics design & development**	**71 days**	**Mon 6/19/00**	**Tue 9/26/00**		**72%**
34		**Attitude/revert panel redesign**	**16 days**	**Tue 9/5/00**	**Tue 9/26/00**		**52%**
35	✓	Mechanical specification	2 days	Mon 9/11/00	Tue 9/12/00		100%
36	✓	Procure	2 wks	Wed 9/13/00	Tue 9/26/00	35	100%
37		**Validate**	**11 days**	**Tue 9/5/00**	**Tue 9/19/00**		**0%**
38		Tester design and fabricate	2 wks	Tue 9/5/00	Mon 9/18/00		0%
39		Evaluate panel	1 day	Tue 9/19/00	Tue 9/19/00	38	0%
40		**RDR-1E/F**	**63 days**	**Mon 6/19/00**	**Thu 9/14/00**		**87%**
41	✓	Electrical design changes	1 wk	Mon 6/19/00	Fri 6/23/00		100%
42	✓	Incorporate lightning/EMI mods	5 days	Mon 6/26/00	Fri 6/30/00	41	100%
43	✓	Procure PWB	3 wks	Wed 7/5/00	Tue 7/25/00	42	100%
44	✓	Build CCA	2 days	Wed 8/23/00	Thu 8/24/00		100%
45		Verification and test	1 day	Mon 9/11/00	Mon 9/11/00	44	0%
46		Integrate four CCAs in chassis	3 days	Tue 9/12/00	Thu 9/14/00	45	0%
47		**Software requirements documentation**	**235 days**	**Tue 2/1/00**	**Tue 12/26/00**		**27%**

Figure 5.3 Tracking Gantt chart.

department staffing plans to accomplish the work, estimate the costs, and kick-off, monitor, and close-out the program.

Table 5.1 outlines the five functions—a description of the function, and the roles of the program manager, department manager, director of product development, and project team.

Schedule control

Product specification changes that impact schedules are approved by the program manager and departments involved. The program manager initiates two kinds of changes—schedule updates based on tracking information, such as percent complete, and more fundamental changes from customer inputs, design change notices, and other more substantial changes in the scope of work. Affected department managers and the appropriate program manager must agree to all

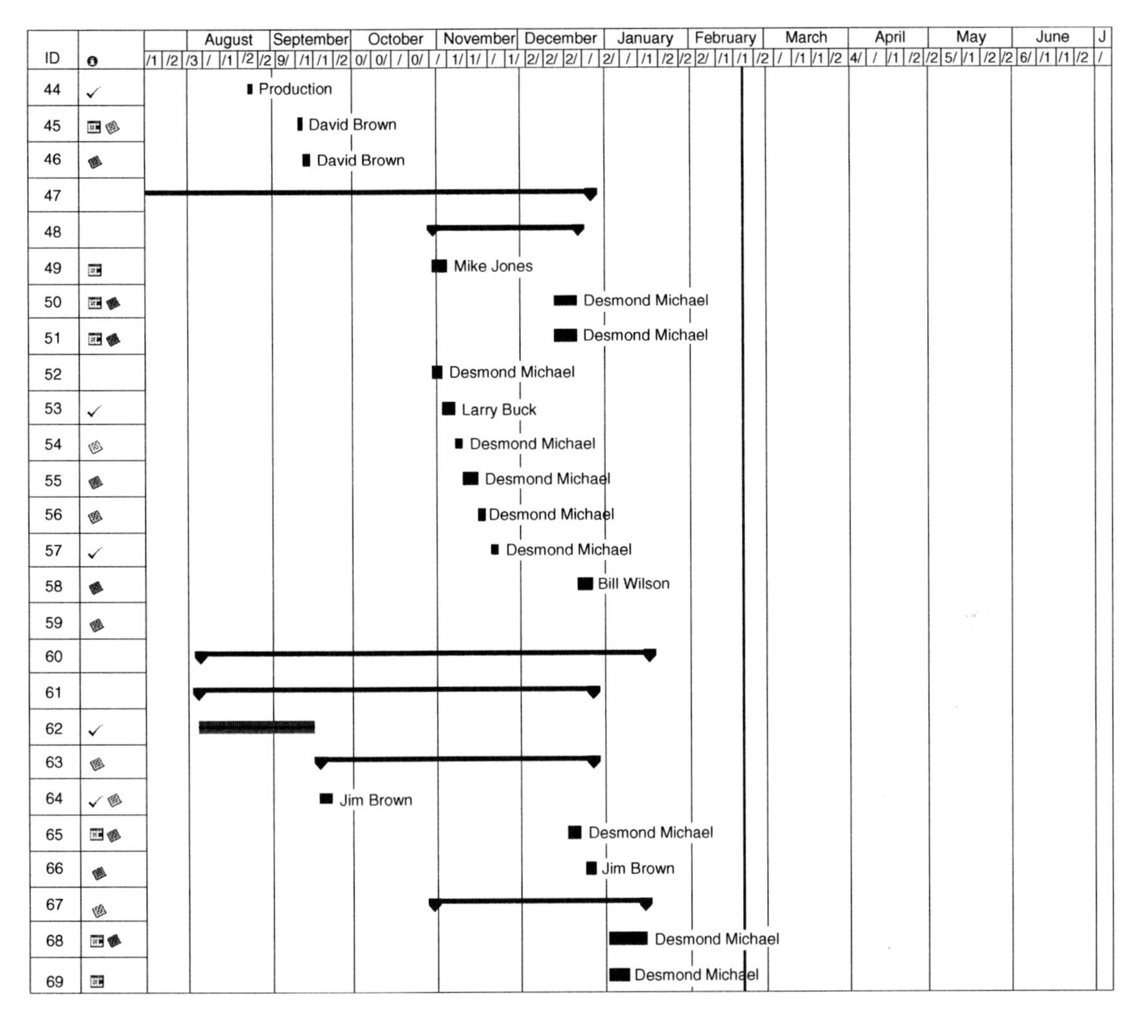

Figure 5.4 Tracking Gantt chart with bars.

schedule changes. After the baseline schedule is saved, the director of product development must approve any slips or changes in scheduled milestones on the critical path and review all risks in monthly program reviews.

Baselining the schedule

After a preliminary schedule is prepared and analyzed, cross program resource conflicts resolved, and successful delivery of the product within the schedule determined to be feasible by the program manager, the schedule is baselined. Establishing the baseline schedule is a significant action in the program management process, signifying the official kickoff of the work and indicating a

TABLE 5.1 The Scheduling Process

Scheduling function	Description of function	Program manager role	Department manager role	Director of product development role	Project team role
1. Develop top-level work breakdown in Microsoft Project or Excel	Develop top-level program structure consistent with product development process	Lead role, working with systems engineer and department management Identify risks	Participates in developing top-level structure Confirm risks	Review, comment, approval	Helps program manager develop structure, as requested
2. Flesh out schedule, establishing task and subtasks, durations, risks, interdependency, and constraints	Identify tasks and risks; prepare risk matrix, prepare preliminary schedule	Lead role with inputs from department staff; works to find ways to accelerate work in concurrent tasks when feasible and address risks	Helps define how scheduled work and inherent risks can take advantage of prior work, supports concurrency in work schedules	Review, comment, approval	Helps department manager find efficient ways to work in parallel
3. Assign resources to schedule, estimating hourly requirements for each task	Establishes resource needs to support schedule; assigns work from scheduled tasks to staff; confirms resource availability	Works with departments to identify program team and meet other resource requirements	Lead role, department managers are responsible for staffing program with competent, adequately trained personnel and providing adequate resources	Review, comment	Participates with department manager and program manager in fitting assignments into current and projected workload
4. Establish risk-based schedule baseline, confirming program "kickoff"	Save schedule as "baseline" in Microsoft Project software, placing file in central directory on network	Takes lead to kickoff program, handing out hardcopy of baseline schedule at meeting, along with resource usage table that defines resource commitments	Participates in kickoff meeting, supports resource usage plan	Approval of baseline before it is saved as such	Review, comment

TABLE 5.1 The Scheduling Process (*Continued*)

Scheduling function	Description of function	Program manager role	Department manager role	Director of product development role	Project team role
5. Monitor performance against baseline, report on performance, variances, and cost to complete, manage change	Enter actual data on percent complete and/or cost data from time-sheet project codes; revises start and finish dates as appropriate	Gets percent complete and other performance information from project team members; report weekly to director of product development	Reviews actual and planned BCWP, and makes recommendations	Reviews actual and planned	Recommends corrective action to offset schedule slippage

strong commitment to the schedule and resource plan. The baseline is the point of departure for monitoring and tracking process. When a project is baselined, the project schedule is complete. Here are some rules of thumb for baselining:

1. The purpose is to get to a baseline schedule that captures all the work to be done. This includes key documentation and procurement tasks. The baseline schedule does not change unless the basic scope changes. Once agreement is reached, the program manager confirms the baseline by saving it and making it available on the network *as the baseline*. There is no uncertainty where the baseline is and how to access it.
2. The baseline schedule is the agreed-upon, scrubbed schedule for the program, linked to the resource pool. The baseline shows all interdependencies, linkages, and resource requirements, includes all tasks necessary to get the work done, and shows impacts on parallel programs and resources. All procurements and test equipment are covered in the schedule.
3. The baseline schedule is resource-leveled—the schedule can be implemented with current, available resources. Assigned staff are aware of the commitments and have "signed-on" to complete their tasks to meet the schedule milestones.
4. Getting to the schedule baseline involves collaboration between the program manager and all departments and staff involved in planning and implementing the schedule. A baseline meeting is held to arrive at a final agreement on schedule and resources committed before the baseline is saved to the network. The program manager facilitates the meeting, and all department managers attend and come prepared to commit their resources to the final, agreed upon baseline schedule.
5. The final review of the schedule at the baseline meeting involves reviewing all stages and tasks, linkages, and resources assigned line-by-line.

6. The baseline schedule is monitored weekly, with actual percent complete data and changes in start and finish dates entered weekly and reported at program review.
7. Risk contingencies from the risk matrix data are plugged into the schedule and durations estimated along with other tasks. A risk-based schedule is prepared.

Baseline procedures

The program manager uses Microsoft's Planning Wizard (or Tools, Tracking, Save Baseline) to set a baseline schedule. At the same time that a baseline is created, a backup copy of the project file is created as a permanent archive of the original schedule for later reference and comparison to actuals.

Sometimes it is necessary to create a baseline schedule before the complete schedule and task structure is determined, simply to serve as a basis for capturing actual progress. This is because some work that is clearly on the critical path, such as mechanical design or long lead-time procurement, must begin immediately to meet key milestones, sometimes before all the details of a schedule are worked out. To accommodate to this, when the final schedule is completed the baseline can be updated by saving an "interim plan." The interim plan saves particular start and finish date changes that are made after a baseline has been saved. Interim changes can be made for the entire project or for selected tasks.

Managing schedules on the network

The basic objectives of network management of program schedules are to: (1) enable the program management department to control schedule updates and schedule versions and (2) provide department managers and staff with an easy way to review and provide input to schedules and schedule assumptions. Here are the steps involved:

1. All schedules will be housed on the server in individual program manager folders.
2. The central resource pool file will be housed in the schedule folder. Archive versions of schedules will be housed in a separate folder.
3. The program management department will control access to schedule files. The director of product development and the program management department (program managers and program administrator/planner) will have "write" access to the schedules. Department managers, systems engineers, and team staff and other users will have "read" access to program schedules.
4. The program management department is responsible for maintaining and updating program schedules on the network. Once the director of product development, the program manager, and department managers agree on a proposed schedule and/or update, the schedule will be linked to the resource

file and resource conflicts will be identified and resolved. The program manager will then save the schedule as a baseline schedule. Once baselined, the schedule will be placed in a designated directory. The baseline schedule will be the only version of that schedule housed on the network (except for archives) and will serve as the source of "planned versus actual" tracking information.

Resource planning and control

Scheduling is essentially the process of planning for use of personnel and equipment resources. Good program management requires that there be a process to plan for the acquisition of future resources, to allocate current and projected resources to schedules, and to make shifts in resource management as required. The process provides for a central resource pool to identify impacts of project schedules and ensure the efficient utilization of the workforce. The resource pool information on the network is shared with management staff and all team members to allow each team member to evaluate the scheduled work assigned and to provide guidance on task definition, durations, start and finish dates, and interdependencies.

While work actually done on a project is tracked automatically by Project when percent complete data are entered, the program management and finance departments collect actual work done from time sheets to gain a more accurate assessment of actual work. Actual work is tracked from time sheets, which collect hours of work against the project account-numbering scheme. The program manager is responsible for establishing the account numbers for charges to the project and for assuring that time sheets are kept for all work on the project. The project administrator/planner is responsible for working with the finance department to collect data each week and enter these into appropriate schedules.

Tracking and program review

The director of product development will hold weekly program-review meetings to discuss broad program issues; detailed technical, resource, and schedule problems; project team performance; and risk mitigation. Program managers are responsible for preparing presentations for these reviews and identifying key agenda items. Department managers and team members attend program review meetings as appropriate.

The program manager tracks the progress of the program on an ongoing basis and updates the schedule on a weekly basis. The program plan is to be updated as changes in plans warrant. Program managers hold periodic reviews with the program team in preparation for reporting and to support task assignments and feedback, either as a single meeting with all functions represented, or as a series of meetings with major functional areas represented at each meeting. Program managers report progress for their portions of the weekly report, due to the director of product development.

Schedule update procedures

Using Microsoft Project, program managers can track and update actual performance information including percent complete for each task, change in task start date, change in task finish date, task duration, task cost, and total work once a project is underway.

1. *Percent complete on a task.* At the very minimum, the program manager updates percent complete for each task on which work has been done during the past week. These data are gathered from the appropriate project team members based on their assessment of the percentage of work actually done compared to the baseline work definition. Program managers are responsible for briefing team members on the importance of accurate assessments of percent complete and how their estimates are used to update project performance. When percent complete is entered, Project changes the actual start date to match the scheduled start date and calculates the actual duration and remaining duration.
2. Add a new task to a baseline or interim plan.
3. Enter actual duration of a task.
4. Enter actual start and finish dates for a task.
5. Enter actual work (e.g., hours) completed on each task (from time sheets).
6. Reschedule incomplete work.

Analyzing variance

The program manager uses tracking data to determine how the actual progress of the scheduled work compares to the original baseline schedule, but more importantly to determine impacts of actual work done on the overall schedule and on critical milestones.

Variances that are tracked include:

1. *Tasks that are starting or finishing late.* Along with updating the task start and finish dates through the Tools/Tracking command, the program manager identifies impacts of the change on linked tasks.
2. *Tasks that require more or less work than scheduled.* Changed durations and/or additional resources can be assigned to tasks that are running late. Project will shift durations and start and finish dates automatically when resource units are changed.
3. *Tasks that are progressing more slowly than planned.* Tasks on the critical path that are progressing more slowly than planned must be addressed either through redefining the task to stay within the schedule or adding resources. Tasks off the critical path provide the program manager with some slack time or "safety buffer" before they go onto the critical path because of delays.
4. *Resources that aren't working for scheduled hours.* If there are major variances in actual work versus planned work, this can have implications for

several projects. Based on actual hours worked, program managers reassign resources and link to the resource pool to broaden impacts and conflicts. Project's resource leveling capability can be used to address resource conflicts as well.

5. *Earned value.* Earned value is an indicator of cost and schedule variance. It is important to track both, whether the actual work is on schedule *and* whether the actual resources expended on the work are consistent with what it should have cost to get the work done based on the baseline budget. In other words, earned value tracks whether the work accomplished actually cost what it should have cost, given the original budget. Reports on earned value show schedule variance and cost variance.

6. *Corrective action.* The real issue in variance analysis and earned value is what corrective action the program manager takes to put a program, which is showing substantial schedule variance, back on schedule. Program managers are responsible for alerting the team to these variances, reporting them in program review, and coming up with corrective actions. Some corrective actions include:
   - Making sure there is no scope creep, that is, the work the team is doing is not on the system requirements specification
   - Implement risk contingency plans
   - Change task dependencies and linkages
   - Assign overtime work
   - Hire or assign additional resources
   - Decrease amount of work necessary to do a task
   - Reassign resources
   - Delay selected tasks
   - Change working hours

   In support of tracking and program review, the program administrator/planner:

- Serves as a resource for Microsoft Project procedures and training
- At the request of a program manager, tracks progress against the schedule and anticipates future schedule problems
- Flags current and new issues for the week from current schedules
- Distributes assignments in the central resource pool to project team staff and gathers feedback
- Identifies conflicts and facilitates resolution

Program closeout and lessons learned

Each program manager meets with the program team at the end of a program to go over the project, identify uncompleted documentation or other tasks, and to identify lessons learned. Lessons learned are captured and reported back to the director of product development and the program administrator/planner for follow-up action.

Defining the Program Management Process and Risk

It is important to describe program manager responsibilities, especially in risk planning and control.

Responsibilities

It is the responsibility of the director of product development to provide leadership and direction to the program management department and department managers in carrying out this policy. It is the responsibility of the program management department to carry out this policy in collaboration with department managers.

The program manager (PM) is responsible for all aspects of a program/project including cost, schedule, and technical performance. He or she has the responsibility for planning, scheduling, tracking, and coordination of a product or series of products and is responsible for maintaining the program management data (schedules, plans, action items, and the like) and ensuring that all technical data are updated as the design and plans change.

The PM manages programs through all stages of product development and along with the systems engineer ensures that *development design* and *risk reviews* are conducted and documented and all actions are resolved.

The program manager:

- Defines clear objectives for each respective program and ensures that all program personnel are informed as to these objectives
- Establishes a program plan that meets all the objectives identified for a given project
- Creates a detailed schedule, in conjunction with team members and department managers, that meets all program objectives
- Provides direction to all team members and any/all departments for the purpose of meeting program objectives
- Tracks progress of a project or series of projects
- Reports progress against the plan and schedule to management
- Identifies and resolves problems associated with the program/project
- Identifies and mitigates developmental risks
- Identifies all hardware needs for each program and ensures that these are included in *forecast management plans*
- Serves as the customer as the primary *point of contact* (POC) for any hardware returns.

Process

The following sections describe the requirements for program management.

Program coordination

The PM is responsible for the overall management and coordination of the program/project during each stage of development. In order to accomplish this the PM is required to conduct periodic project review meetings with the program team (including design, certification manufacturing, and procurement personnel) to ensure a unified and informed effort. The PM is responsible for ensuring that corporate management approves any change in requirements. As program requirements change, the PM ensures that the program plan and schedule and other applicable program data are kept current and distributed.

From initiation of program activities through completion of the product the PM ensures that all program requirements are identified, implemented and verified, all changes are tracked and incorporated, and all data artifacts are produced and maintained through all stages of the product development and production. Along with the engineering organization the PM is responsible for determining the proper effectivity for any design change to products that are in either the product development or production stage.

In the event of issues and resource conflicts that cannot be resolved by the program manager and department managers, the program manager is responsible for raising them to the level of the director of product development for problem resolution and decision.

Program plan

The program manager has the primary responsibility for creating a program plan and schedule of major milestones that identify all program objectives for a given project or series of projects. The program plan will be created based on inputs from the program team members and functional managers. The PM is required to keep the program plan current and incorporate changes as required. The program plan must include, as a minimum:

- Overview of customer requirements, program scope, and objectives
- Identification of major tasks, risks, and schedule milestones
- Basis of program (new development, modify existing product, and the like) and strategy to meet objectives (e.g., qualification by similarity or perform qualification testing)
- Identification of quantity of units needed for development activities and their uses (how many test assets are needed and why)
- Identification of test equipment and components needed for the program (special test software, special test sets, fixtures, jigs, cables, and the like)
- Identification of any special tests required for the program

- Procurement requirements for development efforts
- Manufacturing requirements for development efforts
- Outside test facilities needed
- Outside integration (A/C, customer lab, and the like)
- Estimate of ODC required for the project including materials for test assets, test equipment, outside facilities, travel and any other pertinent costs
- Risk assessment and risk mitigation plans

Program schedule

The PM has the primary responsibility for creating and maintaining a detailed program schedule that meets all program objectives. The schedule must contain, as a minimum:

- Tasks and milestones that correspond to all major program objectives and milestones contained in the program plan
- All tasks required to execute a given program including systems design, detailed design, certification, test equipment, reliability, safety, design reviews, manufacturing, procurement, and test assets
- Tasks detailed to the lowest practical level. Tasks should generally be identifiable to a single resource
- Resources assigned to tasks and leveled reflect a realistic workload
- Identification of manpower requirements for the program

The PM department is responsible for reviewing the resource allocations of all program schedules and anticipating resource conflicts and problems. This is accomplished by assessing each proposed program schedule in terms of its impact on current schedule resource commitments and recommending solutions.

It is the PM's responsibility to identify and present to the respective department manager(s) any schedule and/or resource conflicts that prevent the program from meeting its objectives/goals. If the conflict cannot be resolved with the department manager without impacting these objectives/goals, the PM shall present the conflict(s) to the director of product development for direction. If it is deemed that the conflict still cannot be resolved without impacting the program objectives/goals, it is the PM's responsibility to communicate the problem and impact back to the customer.

Program baseline and kickoff

When effort on a new project/program or phase is initiated, the PM is required to convene a kickoff meeting for the work being initiated. The kickoff meeting should include department staff and product development, manufacturing, quality, and purchasing personnel, as a minimum. The purpose of the kickoff meeting is to present the program plan and schedule and formally initiate program activity.

Program tracking and reporting

The director of product development holds weekly program review meetings to discuss broad program issues, detailed technical, resource, and schedule problems, project team performance, and risks. Program managers are responsible for preparing for these reviews and anticipating key agenda items.

The PM is required to track the progress of the program on an ongoing basis and update the program schedule on a weekly basis. The program plan is to be updated as changes in plans warrant. Program managers are required to hold periodic reviews with the program team, either as a single meeting with all functions represented or as a series of meetings with major functional areas represented at each meeting.

Problems that significantly impact cost, schedule, or product performance must be identified, investigated, and reported to management so that decisions can be made on a fully informed and timely basis. As part of the program tracking and reporting, it is the program manager's responsibility to identify those problem areas and to assume responsibility for resolving them.

Chapter

6

Risk Matrix Samples

This chapter deals with forms and templates for risk matrix analysis.

Steps in Preparing a Risk Matrix

Figure 6.1 diagrams the steps for preparing a risk matrix.

Step 1: Identify tasks with risk. The overall project risk is the sum of the individual risks associated with product development plus the risk associated with the market for the product. In other words, the risks associated with producing the product to specification is different from the risk associated with its market performance. Thus both risks must be incorporated into the risk identification process. The first risk—the project market risk—is identified in business planning and project selection for the company's portfolio. The second risk—the product performance risk—is identified as the total impact of all the task risks associated with its design and production. It is the second performance risk that is identified in this step, task by task. The way this is done involves the work breakdown structure (WBS). Each level and task/component of the WBS is reviewed and ranked in terms of potential risks and then all risks are racked up for the risk matrix exercise. Some risks will disappear when intensities are dimensioned; others will be ranked high and addressed with contingency planning.

Step 2: Describe risk. The description of the risk is a statement that covers what could go wrong with the task. For instance, if the task is product design, the risk could well be the potential for the design to be unworkable when it comes to prototyping and production. In other words, the product cannot be reproduced.

Step 3: Determine impact. The impact of the risk is the change that would occur in key project indicators when such a risk occurs. For instance, the impact of a

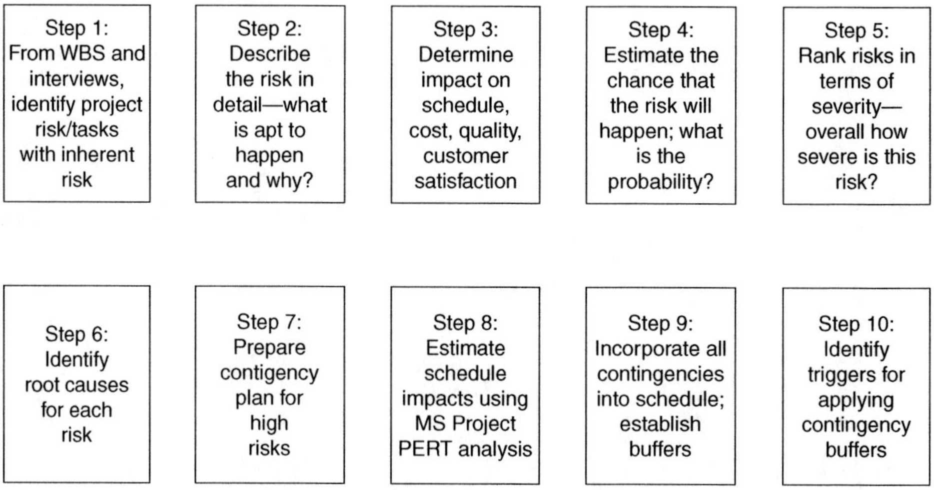

Step 1: From WBS and interviews, identify project risk/tasks with inherent risk

Step 2: Describe the risk in detail—what is apt to happen and why?

Step 3: Determine impact on schedule, cost, quality, customer satisfaction

Step 4: Estimate the chance that the risk will happen; what is the probability?

Step 5: Rank risks in terms of severity—overall how severe is this risk?

Step 6: Identify root causes for each risk

Step 7: Prepare contigency plan for high risks

Step 8: Estimate schedule impacts using MS Project PERT analysis

Step 9: Incorporate all contingencies into schedule; establish buffers

Step 10: Identify triggers for applying contingency buffers

Figure 6.1 Steps in preparing a risk matrix.

design problem would be delayed schedule, increased cost, loss of customer support, and the business risk associated with a failed project.

Step 4: Estimate chance/probability of risk event. The estimation of risk involves ranking a risk in terms of 25, 50, or 75 percent chance of occurrence. Going further with quantification of risk is usually ineffective for most projects because the margin of error in such calculations far outweighs the benefits of going through the mathematics of probability. Project managers are urged to rely on key stakeholder views of risk rankings and past history of similar projects.

Step 5: Rank risk by severity. Here the risk is ranked in terms of severity. Although a risk may have a high level of probability, its severity may be minimal. Thus the cost of contingency is low. On the other hand, a low-probability risk may have a very high severity, e.g., the chance that a key, sole source supplier would go out of business may be low, but the impact would be severe, were it to happen.

Step 6: Identify root causes. Here is the analytic exercise that can actually help manage risks. This involves identifying root causes of anticipated risks and incorporating response to root causes in schedules and task definitions. For instance, if a design flaw is apt to be created by new software then the root cause—the lack of good software suited to the design task—must be addressed with a contingency.

Step 7: Prepare contingency plan. A contingency plan is a schedulable task that addresses the likelihood that a linked task will not work. Such a plan is designed to correct an event or action that delays the schedule or impacts the quality of the work. For instance, a contingency for a failed software integration task might be the outsourcing of a parallel activity to complete integration with outside resources.

Step 8: Estimate schedule impacts using MS Project software. Here the project manager applies the theory of constraints. When a predictable resource or other bottleneck is identified in the planning process, which is created by risk, the project manager identifies the worst case (pessimistic) schedule impact using MS Project PERT analysis and establishes a buffer schedule for that contingency.

Step 9: Incorporate risk in schedules and establish buffers. The new pessimistic, risk-based schedule is not baselined into the project, but a buffer of time equal to the difference between the "expected" and "pessimistic" durations is withheld by the project manager for later use. The project manager will "dole out" these time buffers as needed and triggered by a risk event.

Step 10: Identify triggers for applying buffers. It is important to identify what events or indicators will trigger a buffer action based on risk. Sometimes the project team will see symptoms first and trigger action; sometimes the risk itself occurs and triggers a buffer. Anticipating *how* decisions will be made helps to avoid last minute "crisis management" that inevitably leads to further problems.

Here are some examples of risk matrices.

Note that the Risk Matrix in Table 6.1 addresses a systems development project. Note also that the contingency plan for the "systems not compatible" risk is "testing the system by simulation and in integration." This is a good example of a contingency action that would be placed directly into the project baseline schedule. This way the contingency is embedded into the project schedule to offset the potential risk.

Summing Up: Risk Matrix Examples

These examples are illustrative of the application of the risk matrix tool to *risk management*. The key point here is that each contingency action is not simply "noted." It is embedded into the schedule as a normal task in the baseline schedule. A risk matrix is not designed to establish another list of things to do; its purpose is to help plan and schedule the project so that all contingencies are embedded into the project core.

TABLE 6.1 Sample Risk Matrix 1: Systems Development Project

Risk items	Description of risks	Impact (technical, schedule, cost, quality)	Severity (high, medium, low)	Contingency plan	Ranking
Testing	Critical function needed by new system may be overlooked if not tested properly.	Technical	High	Formal testing plan Test specification Test cases Testing schedule Method to log test results	5
Systems not compatible	Not ensuring the new program will be compatible with the old. If not, an entire new system will be needed.	Technical	High	Test compatibility in simulation and during integration.	5
Network downtime	Network goes down while implementing new system.	Technical	Medium	Be prepared for downtime with slack time available.	3
Lack of information	Not enough research was done during the planning and analysis phases.	Quality	Medium	Have the project well researched before it is approved by upper management.	3
Termination, if applicable	Project termination needs to be done early as not to lose money and time.	Cost/quality	Low	Enough research should have been done to terminate the project before it got to far.	1
Document review	Research documents should be reviewed early on to adjust items to terminate in early stage.	Cost/quality	Low	Management to review all documents for continuation permission.	1

TABLE 6.1 Sample Risk Matrix 1: Systems Development Project (*Continued*)

Risk items	Description of risks	Impact (technical, schedule, cost, quality)	Severity (high, medium, low)	Contingency plan	Ranking
Review data	Review data after approval has been given to make sure the data are correct.	Quality	Low	To be done by management	1
	The approach toward the project has to be acceptable.	Quality	Low	Management to review appro-ach plan during research.	2

Table 6.2 shows a risk matrix on a hiring project.

TABLE 6.2 Sample Risk Matrix 2: Hiring Project

Risk items	Description of risks	Impact (technical, schedule, cost, quality)	Severity (high, medium, low)	Contingency plan	Ranking
Survey	Applicants may not score as recommended on the perso-nality profile and then are not eligible to be hired.	This risk impacts cost because of the cost of the survey, and also impacts schedule because it finds more applicants that can "pass" the survey.	Medium	A very thorough first interview will weed out most people who may not receive as recommended on the Reid survey.	1
Unqualified applicants	Applicants that apply for jobs may not have the desirable qualifications.	This risk impacts the quality of the staff you end up with at store opening.	High	The best way to ensure a qualified applicant pool is to engage in active recruiting. I will actively recruit from other Lowe's locations and from competing retailers in the area.	2

(Continued)

TABLE 6.2 Sample Risk Matrix 2: Hiring Project (*Continued*)

Risk items	Description of risks	Impact (technical, schedule, cost, quality)	Severity (high, medium, low)	Contingency plan	Ranking
Trailer permits	We may not have all the required permits to operate the hiring trailer.	This risk impacts the schedule of the project.	Medium	Prior to opening the hiring trailer schedule an appointment with the city inspector to ensure all permits required have been received.	3
Supplies	There is a possibility that we may run out of supplies.	Schedule impact	Medium	We will inventory supplies daily and place our supply orders weekly	4
Drug testing	There may be applicants who do not pass the drug screening.	This risk impacts both cost and schedule. The drug testing is expensive and is reserved for the applicants who are to be hired. When an applicant doesn't pass the drug screen we will have to look for the next qualified applicant for the job.	This risk rarely occurs so it has a low severity rate.	The biggest deterrent to this risk is the explanation of the hiring process to every applicant. By the third interview the applicant will be told three times that they will be required to take a drug test. We also have signs that say we conduct drug testing, that usually deter anyone who may not pass from applying.	5

Note that this risk matrix covers a hiring process, including a drug test. The task "supplies" includes a contingency to check supplies every day to offset the risk of running out of supplies for drug testing. Again, this contingency would be added to the schedule—each day if necessary—to ensure that drug supplies were available.

Chapter 7

A Case in Risk and Microsoft Project

Good risk management requires substantial organizational capacity to handle risk-related information and data and to calculate schedule and cost impacts of various risk events. As projects become more and more complex, risk management requires an effective project management software program and supporting network systems to allow exchange of data and analysis. Microsoft Project provides a good base for project planning and control and for scheduling and costing out risk mitigation actions. But the organization needs a network system with a workable directory so that project teams and stakeholders can access and communicate timely risk data.

In addition to its usefulness in documenting contingency tasks as a part of the baselined schedule, MS Project's PERT (Program Evaluation and Review Technique) analysis allows for estimating alternative scenarios and impacts on schedule and cost. PERT analysis provides a template for placing weights on three scenarios—expected, optimistic, and pessimistic—and estimating durations as well.

This chapter addresses a case study in how to use MS Project in selecting and scheduling a project, and how to manage project risk. The "Huntsville case" tracks a project from selection and early planning through project initiation, and then into project review after 2 months into the work. The project is presented in four parts. Part 1 is a project selection exercise using net present value and a weighted scoring model reflecting risk to rank two competing projects. Once a project is selected, Parts 2, 3, and 4 involve implementing the selected project—learning the MS Project management software, preparing schedules, producing and interpreting reports, and taking corrective action based on risk events and project progress. The project also involves preparing management reports on project progress and corrective action. "Assignments" are made to the reader to get hands-on experience, and an answer is provided to all "assignments" and question.

Part 1: Portfolio Project Selection and Risk

The Seitz Corporation was founded in 1932 by Johann Seitz. The main products of the firm were small to medium plastic bottles and containers, used mainly in the food and dairy industries.

By 1975 the annual sales of the corporation had reached $31 million and the firm enjoyed a dominant market position in the upper Midwest. In 1982, Walter and Teri Sietz—Johann's grand children—assumed the day-to-day operations of the business. Teri Seitz was a somewhat unstructured, but diligent, student of the latest business school theories and decided that in order to meet increased competition, especially from Japanese companies, Seitz Corporate needed to develop a long-range strategic plan. A consultant was hired to do this. The plan reflected much of the philosophy that had enabled the company to succeed in the past, but this was the first time these philosophies had been written down and synthesized into formal goals. The key elements of the strategic plan, ranked according to priority with number 1 being the top priority, are as follows:

1. Double total sales within the next decade.
2. Develop and market new products based on the company's plastics experience.
3. Reduce dependence on equipment suppliers.
4. Be first or second, based on market share in any region.
5. Attain a national presence in the container industry.
6. Increase productivity.

By 1987 sales had increased to $80.5 million. Much of this increase was the result of new markets established in the northeast, southeast, and western parts of the country. New plants were built in Salinas, California, and Harrisburg, Pennsylvania, and as a result of the product efficiencies, reasonable transportation costs, and aggressive marketing Seitz was now the leader in the western market and held a solid second place in the northeast. The firm had maintained its leadership in the original midwest territory.

The corporation is now in the process of selecting major investment programs for the 1997 to1998 period. A total of $3,000,000 remains available for projects during this period. One selection remains and will be made from two candidate projects, which have been culled over from over a dozen project proposals. The details of the two projects are as follows.

Candidate Project 1

Steve Pokorski, the Vice President of Operations, and Joe Downs, the Director of Plant Engineering, have submitted a proposal to replace the extrusion equipment in the West Milwaukee plant. The proposed new equipment is to be manufactured based on a prototype designed and built in Down's developmental facility. All existing equipment had been acquired from three major suppliers. Pokorski authorized the prototype because he felt that the suppliers had become unresponsive to Seitz's bid requests. The technical superiority of the new type of equipment will allow for significant improvements in productivity. The following are the supporting financial data:

Initial investment	$2,475,000
Year 1 Return	$700,000
Year 2 Return	$800,000
Year 3 Return	$800,000
Year 4 Return	$800,000
Year 5 Return	$800,000

Risks. Project 1 initially presents many risks that should be incorporated in the project selection process:

1. Technical risk—new equipment failure
2. Contract risk—supplier delivery failure
3. Performance risk—poor estimates of productivity payoff

In each case the risks are major because they would substantially delay completion of the project and its viability.

Candidate Project 2

Janis Clark, Vice President of Marketing, has proposed another project, which would establish a new plant in Huntsville, Alabama. During the last 3 years sales have been steadily increasing in the area to the point where Seitz has achieved third place in the market share. Clark claims that the market is extremely price sensitive and Seitz's more efficient production capacity would ensure quick attainment of the second position if transportation costs to markets in the southeast were lowered. The financial data to support this proposal are shown below:

Initial Investment	$2,550,000
Year 1 Return	$550,000
Year 2 Return	$600,000
Year 3 Return	$900,000
Year 4 Return	$1,150,000
Year 5 Return	$1,300,000

Risks. Project 2 presents many risks because it is a new construction effort:

1. Construction design failure
2. Design risk—space layout and production process problems
3. Performance risk—unforeseen labor issues
4. Cost isk—inaccurate estimates of cost

Exercise

A. Compare the projects using a weighted scoring model based on the company's strategic goals, risks, and project descriptions.

B. Using a discount rate of 12 percent, calculate the net present value for each of the projects.

C. Make a selection of the most desirable project based on the results of A and B. Support the selection with comparison charts and net present value calculations.

D. Prepare a report to management supporting your decision.

Note: The purpose of this assignment is to illustrate to a project manager and the project team the importance of being involved in the actual selection of the project and knowing its background and assumptions. Analysis of projected cash flows, risks, and net present value provide valuable information on what is expected of a project as well as information that will be useful to the project manager once a project is selected. For instance, net present value requires an estimate of cash flow over the life cycle of the project, which requires some analysis of initial investment costs and expected revenues on a regular basis. Such a cash flow analysis usually requires some planning and outlining of the project to identify key milestones and deliverables which can be invoiced to the customer for progress payment. The weighted model requires assessment and ranking of projects against strategic goals, as well as the determination of actual weights for each. This exercise should provide valuable background and insight as the project manager enters the initial phase of the project, once selected.

Introduction to Parts 2, 3, and 4

Parts 2, 3, and 4 of the course project will give you the opportunity to use the project management software to construct a preliminary schedule (Part 2), "scrub" and adjust the schedule to respond to changes and risks, and add resources (Part 3), and track, interpret, and take corrective action based on actual progress (Part 4). You are provided relevant schedule, cost, and performance data to enter into the software, and you are asked to interpret progress and prepare reports to management on issues such as conflict management and organizational structure. Remember that the software automatically calculates and constructs the Gantt chart and critical path from your data entries, thus bypassing the manual development of system diagrams and calculation of earned value and other indicators.

Entering data into MS Project to reporting on a project is sometimes challenging. Sometimes the data are difficult to enter; sometimes the results are not what you expected. There are no instructions in the course project on the software, so you are expected to use the required software manual to guide your data entry and interpretation. The software may not work exactly as the manual suggests, and you may encounter results and reports that differ from those of your colleagues. All this demonstrates the experiences of real project managers in real project situations.

There is no right answer to this case, even if results such as final budget and earned value indicators should fall in a predicted range. Work on clear, analytic answers to the questions posed, especially on corrective action. (What would you do as a project manager given your interpretation of progress and problems?)

Part 2: Project Planning and Risk

In January 1997 the board of directors of Seitz met in Milwaukee with several important questions on the agenda. Among the decisions to be made was the selection of investment projects to be executed during the next fiscal year. After a detailed presentation, one of the proposed investments, the construction of a new plant in Huntsville, Alabama, was approved. Janis Clark, who had submitted the proposal, was given the responsibility for executing the project. In the memo which informed her of the board's decision she was given authority to spend $2,750,000 and was given a target date of June 1998 for completion of the plant and the first shipment of the product.

Besides being provided with funds in excess of the original budget request, Clark had been given access to other functional elements of the corporation's Midwest plant and headquarters for assistance in the project. Steve Pokorski, the Vice President of Operations, and Joe Downs, Director of Plant Engineering, (who had submitted an alternative proposal which the board had rejected as being "reactionary and backward in thinking") were instructed by the board to provide Clark with whatever she needed, even if it meant that their own performance would suffer. Down's prototype was, of course, turned over to Clark so that she could use the technology in the new plant.

Clark immediately called in her regional sales manager and her marketing director to assist her in initiating the project. They decided that the first objective was to put in place an appropriate organizational structure and to staff it with the required personnel recruited from both internal departments and from outside the company. Being a marketing and sales organization they were a little thin on people with technical skills, but they expected to utilize the best and brightest from Down's and Pokorski's organizations because as Clark said, "They don't really need much talent to run their operations anyway. We can offer their best people higher level management positions on the project team and then at the new Huntsville plant." When one of her staff questioned if these people would all be willing to move to Alabama, even for a promotion, she replied "who in his or her right mind would want to stay in Milwaukee when they could move to Alabama; and with a promotion, too!"

Although they expected the project manager to finalize the plan, Clark and her assistants drafted a preliminary list of the tasks, which they felt were needed to accomplish the project, together with precedence relationships and task durations.

Clark had decided that one of the products to be produced in the new plant would be a new plastic container for wine products. She felt that glass bottles were on their way out as wine containers and that this was a ground floor opportunity to capture a major new market segment. In order to introduce the product properly, however, she felt that the plant would have to be ready no later than the end of 1997.

Assignment. You are a consultant who has been hired by the directors of Seitz Corporation to monitor the initial phases of the project. Answer the following questions in the form of a confidential report to the board on your evaluation of the project to date and recommendations.

1. Based on the information given above, define an appropriate organizational structure for the project. Choose among the matrix, functional, and pure project organizational structures covered in the text. Include in your organizational chart details of how the project should tie-in to the parent company, and how the project itself should be organized. Provide an organizational chart and give the reasons for your selection.
2. Identify any potential conflicts between Clark, Downs, and Pokorski. Identify the type of conflict, the effects they might have on the project, and how they might be resolved.
3. For Parts 2, 3, and 4 of the Course Project, you will be using Microsoft Project to document the project. So study the program manual and get familiar with the software. For this part (Part 2), follow directions to start the project and save the project file first. Set the current date (pull down the project information box from the Project menu to enter dates) as March 17, 1997, and the project start date as April 17, 1997. Then enter task data onto the Gantt chart from Table 7.1 (preliminary plan). Print out the preliminary schedule, a printed Gantt chart.
4. Prepare a risk matrix on the Huntsville project, complete with contingency plans.
5. Do any issues about the project tasks and their relationships "jump out" at you—are the tasks in a logical sequence and does the overall structure and "contour" of the project look right?
6. Will the plant be ready by the end of 1997 as Janis requires? If not, will it meet the June 1998 deadline set by the board of directors? What options might be open to ensure that these deadlines are met, if the current schedule indicates that the plant will not open in time?
7. Prepare the report to the board covering your answers to the above questions.

Notes on the software. Don't forget to change the current and status date of the project for Parts 3 and 4. This is key to the effectiveness of the software in "knowing" what dates you are working with at any time. Be sure to save the file when you are finished, but *don't* save it as a baseline yet. Remember also, in printing out the Gantt chart, to adjust the timescale of the bar chart to a scale that allows you to print the whole project on one page. Go to the Format menu

TABLE 7.1 Huntsville Preliminary Plan

Task number	Task name	Duration (in weeks)	Predecessor
1	Select an architect	2	None
2	Recruit and train managers	6	None
3	Select real estate consultant	2	None
4	Create production plan	4	None
5	Building design	6	1,4
6	Site selection	3	3
7	Select general contractor	2	5
8	Permits and approvals	3	3,6,7
9	Building construction	24	8
10	Plant personnel recruiting	8	2,4
11	Equipment procurement	24	4
12	Raw material procurement	8	4
13	Equipment installation	4	11
14	Product distribution plan	2	4
15	Landscaping	3	9
16	Truck fleet procurement	8	14
17	Preproduction run	4	10,12,13
18	Production start-up	1	17
19	Distribution	1	16,18

and click Timescale, then adjust the major scale and minor scales until you can place the whole project on one page.

Part 3: Establishing the Risk-Based Project Plan Baseline

During the various meetings held by Janis Clark, one of the major priorities was the selection and appointment of a project manager for the new Huntsville plant project. The decision had been made to use a mixed organization, which would provide the project manager with an independent staff and would also allow the use of some resources from the existing functional organization of the corporation. Clark wanted to have the project staff transition into the operational management roles of the new plant upon completion of the project. Clark contacted Steve Pokorski, Vice President of Operations, to seek his assistance in filling the project manager position. Pokorski was interested in assisting because he realized that once the Huntsville plant was completed the responsibility for ongoing operation would be his. At first there was conflict between Clark and Pokorski because of the rejection of Pokorski's alternate proposal, but his had been dealt with by Teri Seitz who, upon recognizing the seriousness of the conflict, and having no desire to let it fester, had hired a management consultant, The SIGMA group, specializing in team building and conflict resolution.

Pokorski had been through similar projects in the establishment of the Salinas plant and the Harrisburg plant and had learned some difficult lessons along the way. The first project (Harrisburg) had been implemented by a project team organized and run out of the corporate office. The staff returned to their original jobs after the project was complete, with the plant operation managed by a team locally recruited during construction. Pokorski had decided after a long and difficult start-up period that in the future the availability of potential managers recruited internally would be a priority.

Pokorski had proposed four candidates who he discussed with Clark and between them they selected two possibilities. On two occasions the candidates were brought to Milwaukee and interviewed. Andy Piltz, the Material Manager in the Salinas plant, was chosen and accepted on March 15, 1997.

Andy took about 2 weeks to transfer his responsibilities in the Salinas plant during which time he made a preliminary organization plan and identified with Pokorski's help some possible project team candidates. By the end of March he had contacted his candidate list and received commitments for the three positions he had decided to fill internally. The staff list included:

Project duty	Plant position	Current job
Production specialist	Manufacturing manager	Senior manufacturing manager
Marketing specialist	Marketing manager	Regional sales manager
Facility specialist	Maintenance manager	Asst. maintenance manager

It was decided that accounting and personnel managers should be recruited locally by the project manager and undergo training in Milwaukee. All other plant project and plant staffing decisions would then be the responsibility of the personnel manager. After several meetings with his staff, Piltz completed the task list (see Table 7.2, revised plan) by adding a summary task for the

TABLE 7.2 Revised Huntsville Plan

Task number	Task name	Duration (in weeks)	Predecessor
1	Huntsville Project		
2	Select an architect	2	
3	Recruit and train managers	6	
4	Select real estate consultant	2	
5	Preproduction plan (new)	2	4
6	Create production plan	4	
7	Building concept (new)	2	2
8	Building design	6	
9	Site selection	3	
10	Select general contractor	2	
11	Permits and approvals	3	
12	Building construction	24	
13	Plant personnel recruiting	8	
14	Equipment procurement	24	
15	Raw material procurement	8	
16	Equipment installation	4	
17	Product distribution plan	2	
18	Landscaping	3	
19	Truck fleet procurement	8	
20	Preproduction run	4	
21	Production start-up	1	
22	Distribution	1	

TABLE 7.3 Huntsville Resource and Fixed Cost List

Resource/ cost initials	Full name	Cost type	Hourly rate ($/hr)	Fixed cost ($)
FS	Facility specialist	Personnel	21.00	
PM	Project manager	Personnel	32.00	
CP	Corp. personnel	Personnel	18.00	
PS	Production specialist	Personnel	24.00	
BD	Building design	Fixed		32,000.00
RC	Real estate consultant	Personnel	55.00	
SC	Site cost	Fixed		216,000.00
ME	Manufacturing engineer	Personnel	18.00	
GC	General contractor	Personnel	70.00	
PC	Permit costs	Fixed		3,200.00
BC	Building costs	Fixed		1,200,000.00
PD	Personnel director	Personnel	20.00	
EC	Equipment costs	Fixed		780,000.00
AC	Architect	Personnel	65.00	
EI	Equipment installation	Fixed		120,000.00
MS	Marketing specialist	Personnel	24.00	
AD	Accounting director	Personnel	22.00	
TC	Truck cost	Fixed		135,000.00
TM	Traffic manager	Personnel	19.00	
LC	Landscaping	Fixed		26,000.00
PA	Purchasing agent	Personnel	22.00	

Huntsville Project, adding task 5, *preproduction plan*, and task 7, *building concept*. At this time the Project team also prepared Table 7.3, resource cost list, and Table 7.4, resource loading list.

Assignment for data entry to Microsoft Project

1. Enter Andy Piltz as Project Manager (go to File and Properties) and set the current date to April 17, 1997 (go to Project and Project Information box).
2. From Table 7.2, insert the two new tasks and task durations into the project task list. You will have to change the predecessors to establish the correct linkages after the two new tasks have been added.
3. Create a summary task for the Huntsville Project—a new task added at the top of the list and "out dented" from the rest of the tasks. (When you have added the summary task at the top of the list of tasks, block the entire list of tasks below and click the out dent button (the one that points to the right) at the upper left of the screen.)
4. Enter resource (both personnel resources and fixed equipment resource) costs:
   - First enter personnel resources. Using the information supplied on Table 7.3, resource cost list, enter personnel resources and their hourly rates by entering them on the Resource Sheet. (Go to View, Resource Sheet, then entry table.)

TABLE 7.4 Huntsville Resource Loading List

Task resource assignments	Resource name–percent assigned				
2	PM – 10%	FS – 60%			
3	PM – 15%	CP – 80%			
4	PM – 10%	FS – 50%			
5	PS – 80%	ME – 85%	PM – 20%		
6	PS – 60%	ME – 80%	MS – 20%		
7	FS – 40%	AC – 70%			
8	BD – 100%	FS – 20%	AC – 10%		
9	RC – 70%	SC – 100%	MS – 20%	FS – 30%	
10	PM – 10%	FS – 25%	PS – 25%	AC – 20%	
11	PM – 10%	RC – 30%	AC – 10%	GC – 15%	PC – 100%
12	PM – 10%	FS – 20%	BC – 100%	GC – 35%	
13	FS – 10%	PS – 10%	PM – 10%	MS – 10%	PD – 80%
14	ME – 40%	PA – 25%	EC – 100%		
15	PS – 10%	ME – 10%	AD – 10%	PA – 20%	
16	PS – 10%	ME – 30%	FS – 40%	EI – 100%	
17	MS – 45%	TM – 40%			
18	FS – 15%	LC – 100%			
19	TM – 20%	AD – 10%	TC – 100%	PA – 10%	
20	PS – 15%	ME – 70%			
21	PS – 30%	ME – 70%			
22		MS – 20%	TM – 60%		

- Then enter fixed costs, using the information supplied on Table 7.3. (Note that fixed costs are entered in a different view and table—the Gantt view and the cost table—consult your manual)

5. Then assign those resources to tasks. Assign hourly resources from the attached Table 7.4, resource loading list. (You assign resources by highlighting each task, then clicking to get the task information box, then clicking resources and entering the resource and the percent assignment to that task). Note that some of the resources you are assigning are fixed costs (like SC, site costs)—they are assigned 100 percent to one task.
6. Then print the Resource Sheet showing resource name, initials, and costs.
7. Print a project summary report and a Gantt chart.
8. Answer the following questions relying on this view and other appropriate reports:
   - Is the Project within budget? What are the budgets, both total and by month, for each task?
   - Are there any resource conflicts (overallocated resources)? If so, what resources are overallocated and for what time periods?
   - What is the current total cost and schedule for the project? How does this compare to the original plan?

- What is the critical path of the project? (Go to Project, Filtered, then click Critical and the Critical Path tasks will be shown on the screen. Print the screen.)
- Your answers to these questions should be in the form of a project start-up report addressed to Janis Clark. It should include a summary analysis of the reports generated by MS Project, which should be attached as backup and referenced in the your analysis.

Since the project plan is now completed, save the project as a baseline schedule. Saving a project as a baseline is a significant action, in effect signaling an authorization to initiate the project work. A project kickoff meeting is typically held to emphasize transition into project implementation.

Part 4: Project Review for Progress, Risks, and Earned Value

Assume now that the project has been kicked off and has been running for 2 months. Andy Piltz is conducting his biweekly project review meeting on June 15, 1997, for the Huntsville Project. He is particularly interested in reviewing earned value of the project so far and determining what corrective action he might take to keep the project on course. The project started on time and information regarding time, progress in terms of percent complete, and expenditures have been collected. The attached table on *actual resources expended* is a summary of costs through June 15, 1997, 2 months into the project. Using data reported from task leaders, the table also presents "percent completed" for each task, and the amount of effort put in by the various resources on their assigned tasks.

Assignment

1. In the Project Information Box (Project, click information box) set your current and status date to June 15, 1997. Do not shut down the computer at this point; proceed to enter actual days worked in the next step.
2. Update your project baseline plan with the actual days the assigned personnel worked on each task (Table 7.5). To enter the actual days go to View, Task Usage, Table, then Work and then enter days into *actual* column.
3. Print an earned value report.
4. Interpret the report, using the data on budgeted cost of work schedule (BCWS), budgeted cost of work performance (BCWP), actual cost of work performance (ACWP), and schedule and cost variance.
5. What is the new critical path of the project? (Go to Project, filter, critical.)
6. What corrective actions are suggested, given the results of this project on June 15, 1997? Prepare a status report to management, which clearly conveys the status of the project and your recommendations for corrective action. Support your recommendations with appropriate reports from MS Project.

TABLE 7.5 Actual Huntsville Resources Expended (as of June 15, 1997)

Task	Resource name–days			
2	FS 6 days	PM 1 days		
3	CP 32 days	PM 4.5 days		
4	PM 3 days	FS 4 days		
5	ME 8.5 days	PM 4 days	PS 8 days	
6	ME 16 days	MS 4 days	PS 12 days	
7	AC 10 days	FS 4 days		
8	FS 4 days			
9	MS 3 days	RC 12 days	FS 6 days	
10	AC 2 days	FS 2.5 days	PM 1dy	PS 2.5 days
11	AC 1.5 days	GC 2 days	PM 1.5 days	RC 4.5 days
12				
13	FS 4 days	MS 2.5 days	PM 4 days	PS 4 days
14	PA 1 dy	ME 8.5 days		
15	AD 4 days	ME 4 days	PA 4 days	PS 4 days
16	MS 6 days	TM 4 days		

Remember that you are being asked to both interpret the indicators *and* come up with corrective actions. Corrective actions might include schedule adjustment, team performance feedback, finance and budget, reviewing percent complete data, looking at quality impacts, and looking ahead to anticipate ways to adjust the project given progress to date.

Suggestions. The earned value reports should show some figures (BCWS, BCWP, and ACWP) in some task boxes for the tasks that started before June 15, 1997—your status date. Sometimes, however, the earned value reports do not show *any* numbers in *any* boxes indicating that the software did not record your actual-days-worked entries. This can happen if you don't change the current and status dates in the project information box before you enter the actual-days-worked information in Part 4. So you will have to correct the data entry so that you have earned value figures to assess. So if you find you have no earned value figures in your report, resave the schedule as an interim plan, reenter the actual days, make sure the current and status dates are entered as June 15, 1997, and print the earned value report, which should now show figures for those tasks that started by June 15.

Risk and This Case

This case is exemplary of a complex business problem of project selection and implementation that addresses risk at every turn. Risk was apparent in how the two projects were evaluated and selected; the company tries to balance risk with payoff. Cash flows are subject to risk because they may not be backed up with good research and analysis. Risks inherent in the projects themselves might not be clear since to do a thorough risk matrix each project WBS would have to be completed and risks identified and ranked in a risk matrix. Cost estimates did not reflect those factors.

Once the Huntsville project was chosen, its risks were addressed in initial scheduling and estimating project resource and costs. Risks were inherent in the earned value estimates in Part 4 in the sense that the estimates of actual resources and associated percent complete were not accurate.

The Solution

In the selection process we would propose that Project 2 be approved. This project will allow the company to meet most of its goals that have been listed as priority—double total sales within the next decade, develop and market new products, reach first or second position in regional market shares, attain a national presence in the container industry, and increase productivity. All objectives will be accomplished, below the remaining project budget of $3,000,000 coming in at $2,550,000 from start to finish. Based on the calculated analysis from the *net present value* and *undiscounted payback,* the payback period will be less than 4 years, leaving 6 years within the 10-year range to double sales. A team will need to be dedicated to the project to address all the variables but problems are not expected because we have gone through this several times already. This makes the planning and implementation processes of the project smoother because our technical support is familiar with the procedures. This is the ideal time to start this project because the market is yearning for our products. Reports show that the market is particularly price-sensitive in the Huntsville area giving our economical products an advantage. Compared to Project 1, Project 2 is the preferred project to put into operations looking at the calculations from all three sources. Below you will notice two models used in scoring the projects—weighted scoring model analysis (Table 7.6) and the net present value analysis (Table 7.7):

TABLE 7.6 Huntsville Weighted Scoring Model Analysis

Weighted Scoring Model				
			Projects	
No.	Category	Weight	One	Two
1	Double total sales within the next decade	25	5	5
2	Develop and market new products based on the company's plastic experience	20	2	4
3	Reduce dependence on equipment suppliers	20	5	2
4	Be 1st or 2nd, based on market share in any region	15	2	5
5	Attain a national presence in the container industry	10	2	4
6	Increase productivity	10	4	4
	Total weighted score	100	280	375

TABLE 7.7 Huntsville NPV Analysis

Year	Cash flow	Rate	NPV
		Project 1	
0	–$2,475,000	1.000000	–$2,475,000
1	$700,000	0.892857	$625,000
2	$800,000	0.797194	$637,755
3	$800,000	0.711780	$569,424
4	$800,000	0.635518	$508,414
5	$800,000	0.567427	$453,942
			$319,535
		Project 2	
0	–$2,550,000	1.000000	–$2,550,000
1	$550,000	0.892857	$491,071
2	$600,000	0.797194	$478,316
3	$900,000	0.711780	$640,602
4	$1,150,000	0.635518	$730,846
5	$1,300,000	0.567427	$737,655
			$528,491

In evaluating the two projects we used a comparative study that was based on how well the two projects met their strategic goals. To set up the weighted scoring model we first assigned a weight or percentage that represented the importance of the goal. Next, a score was assigned to every category for each project, where the score would represent how well each project met its objective. For clarification purposes, a score of 1 represents the lowest score and a score of 5 represents the highest score. The percentages are particularly important because Seitz Corporation has prioritized their strong philosophies. These philosophies are part of the strategic plan to achieve their long-term goals. Rationality for each score relates to the contribution that each category gives to the organization meeting its goals. In the category of doubled total sales, project 2 would presumably achieve those goals quicker because the perspective location is already showing evidence of increased sales. Sales that are steadily increasing yield evidence of prominent possibilities. To add to the ideal marketplace, Clark claims that the "market is extremely price-sensitive," which gives Seitz's economical products an advantage over competitors thus allowing them to attain second position in the southeast market. This is particularly important because aggressive marketing would not be needed in Project 2—Seitz would be providing the region with products consumers are asking for through their purchasing tactics. Thus Seitz Corporation would get a quicker return on their investment not having to divert money to boost sales. The slow or slack initial period often seen when manufacturers move to a new area would not be seen—that period when natives to the area are observing quality and prices of the new product comparing it to their usual purchase, which they are accustomed to.

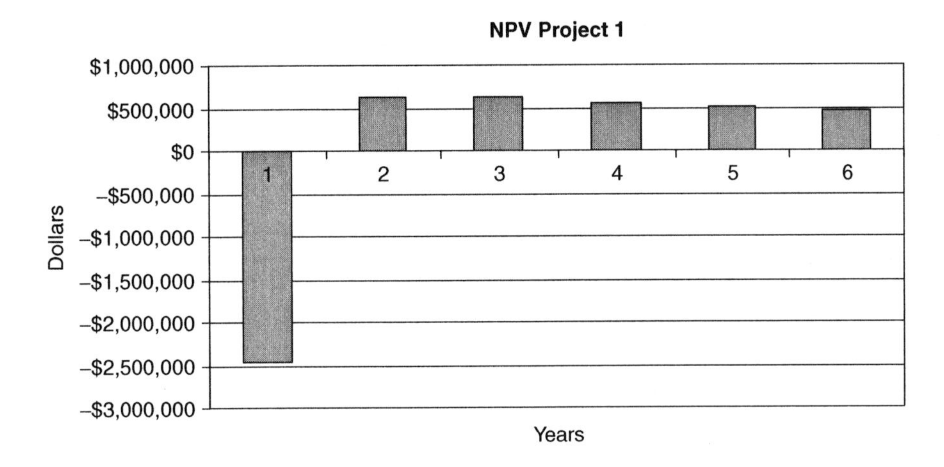

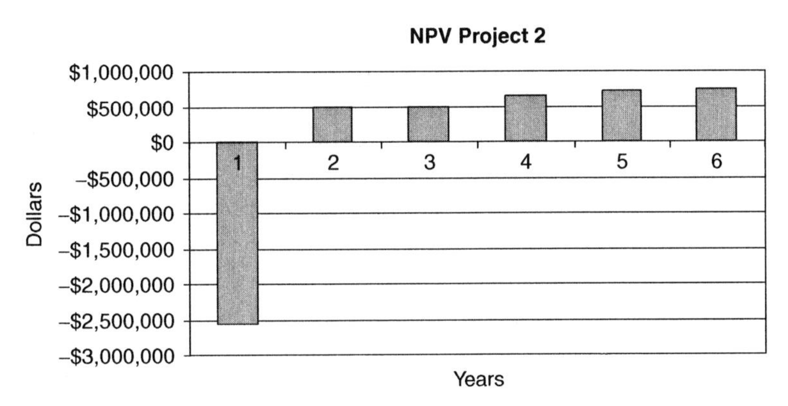

Figure 7.1 Huntsville NPV chart.

This would also reduce project longevity, enabling participants to return to functional projects, and reduce integration of functional groups because once the machines are installed and are operating properly manufacturing will begin and the plant manager will oversee operations.

As a result, we propose Project 2 be selected. The following information will also show why Project 2 seems to be more in line with the company's strategic goals and objectives (Fig. 7.1).

Organizational Structure

Based on the information gathered, our recommendation as to the type of organizational structure should be the pure project organization. The project is separated from the rest of the parent system. It becomes a self-contained unit with

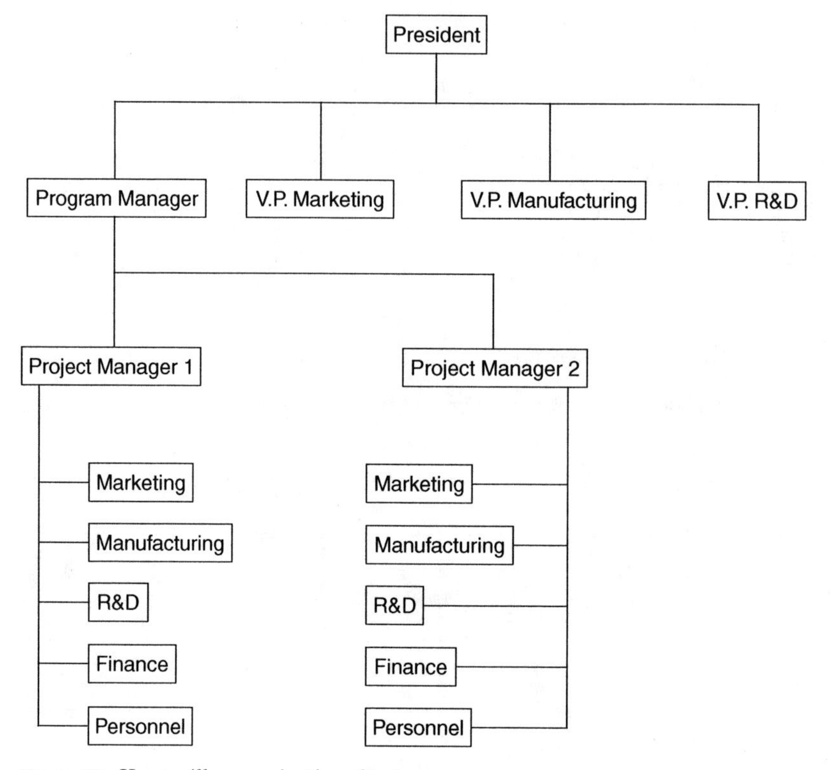

Figure 7.2 Huntsville organization chart.

its own technical, marketing, finance, research, personnel, and development staff tied to the parent firm by the tenuous strands of periodic progress reports and oversight. We would suggest the following personnel be placed in each department (Fig. 7.2):

Marketing/Sales: Janis Clark

Manufacturing: Steve Pokorski

R&D: Joe Downs

Finance Outsource: hire from outside

Personnel Outsource: hire from outside

We chose Janis Clark for the marketing/sales position because that is where her talent will be best. We chose Joe Downs to lead research and development because of his role as the director of plant engineering we felt that position would suit him best. Steve Pokorski was given the position of manufacturing because of his extensive knowledge in operations. By utilizing this particular structure, the lines of communication are greatly lessened, which often leads to better communication between departments. Because authority is centralized, the ability to make quick and accurate decisions is greatly increased. The positions marked outsource (hire from outside) will allow the company to hire local talent to round out the team without having to draw personnel from other parts of the company.

Possible Conflict and Resolution

As for areas of conflict, Pokorski and Downs may have negative feelings toward Clark due to their rejected proposal in favor of Clark's. Another area of conflict could be that Clark's request of resources from Pokorski and Downs may cause their departments to perform poorly even though they are instructed by the board to provide Clark with whatever she needs, even if it meant that their own performance would suffer. Next, we would definitely feel some contention due to the fact they may disagree with Clark every step of the way because of their opposite version of the proposals. Last but not least is the fact that Clark's project will overlap with Pokorski and Downs' responsibilities creating more rocky roads. Recognizing the seriousness of the potential conflict the company will need to have a team building and conflict resolution retreat headed by the management consultant, the SIGMA group, including Pokorski and Downs as part of the project team. Including Pokorski and Downs will give them a sense of ownership as part of the team.

Conflicting Schedule

Do any issues about the project tasks and their relationships "jump out" at you—are the tasks in a logical sequence and does the overall structure and "contour" of the project look right? At first glance project tasks and their relationships seem to run smooth until you really start looking at some areas of conflict. For instance, the plant personnel being recruited well in advance of production—what will they be doing and doesn't this drive up cost. The next thing would be having the raw materials delivered before the building has actually been constructed. Equipment installation is to be completed before the building is actually constructed. That means, the schedule will have to be changed. As for production start-up, we would think you would need a building constructed in order to begin distribution.

Plant Readiness

As for the plant, being ready by the end of 1997 is highly unlikely. As of today it is scheduled to be completed on January 14, 1998, which meets the deadline set by the board of directors. It is possible for the plant to be ready by Janis deadline if the building can be constructed earlier by at least 1 month.

Project Cost

According to my figures, we show this project to be within budget. The total project cost is $2,612,970.00. Please see the following chart:

1. Huntsville Project	$2,612,970.00
2. Select an architect	$1,264.00
3. Recruit and train managers	$4,608.00
4. Select real estate consultant	$1,096.00
5. Preproduction plan	$3,272.00
6. Create production plan	$5,376.00
7. Building concept	$4,312.00
8. Building design	$34,568.00
9. Site selection	$221,952.00
10. Select general contractor	$2,196.00
11. Permits and approvals	$7,604.00
12. Building construction	$1,230,624.00
13. Plant personnel recruiting	$8,352.00
14. Equipment procurement	$792,192.00
15. Raw material procurement	$3,456.00
16. Equipment installation	$122,592.00
17. Product distribution plan	$1,472.00
18. Landscaping	$26,378.00
19. Truck fleet procurement	$137,624.00
20. Preproduction run	$2,592.00
21. Production start-up	$792.00
22. Distribution	$648.00

Resource conflicts (overallocated resources)

Yes, there are resource conflicts. There are three resources that are overallocated:

Facility specialist is overallocated the weeks of 4/13/97 to 4/27/97 by 30 percent.

Production specialist is overallocated the weeks of 4/13/97 to 4/27/97 by 120 percent.

Manufacturing engineering is overallocated the weeks of 4/13/97 to 4/27/97 by 195 percent.

Current cost and schedule

The total cost is $2,612,970.00 and the project end date is 2/18/98. If we followed the original plan according to the board of directors, where the allocated budget was $2,700,000, and completed by June 1998 then we would be within budget and ahead of schedule. If we followed the original plan according to Janis Clark, where the allocated budget was $2,550,000, and completed by end of year 1997 we would be over budget and behind schedule.

Critical path

The critical path is 1, 6 (create production design), 7 (building concept), 8 (building design), 10 (select general contractor), 11 (permits and approvals), 12 (building construction), 18 (landscaping).

Earned value analysis

The information represented below will support the earned value interpretation that follows:

BCWP	$170,848.74
BCWS	$448,108.00
ACWP	$1,061,668.80
SV (schedule variance)	($277,259.26) behind schedule
CV (cost variance)	($890,820.06) over budget

My interpretation of the numbers is that we are over budget and behind schedule by a substantial margin. But after taking a closer look at the numbers and the charts you'll notice that we have procured equipment and our truck procurement that was not needed or could have waited, which pushed my variance in the wrong direction.

Critical path

The new critical path for the project is: 4 (select real estate consultant), 11 (permits and approvals), 12 (building construction), and 18 (landscaping).

Corrective action

Corrective actions that will get us back on track will be to make procurements when they are necessary in the time of the project. As for the schedule, if we see after further investigation that we are really behind then we would have to add more resources or extend the schedule out or both. Also, investigate any slippage and determine whether or not the schedule will need some type of adjustment. After checking and verifying reports for accuracy, as discrepancies could go toward explaining the schedule variance, the next thing we would do is go to the accounting department to verify whether the reports match. From there, we would have a meeting with my team members to find out where they are and get any feedback on what we can do to get back on track. Finally, we feel there is a good possibility that the tasks can be rearranged so that the schedule completion may be faster and actually go smoother. We would reanalyze the scheduling of the three positions that were overallocated (facility specialist, manufacturing engineer, and project manager), which played a major role in the project.

Risk Analysis in Project Selection

Risk can be factored into project selection by assigning a risk ranking to each project (Tables 7.8 and 7.9). For instance, recognizing that Project 1 faced technical, contract, and manufacturing performance risk, each risk can be ranked in terms of risk cost, or the cost of implementing key contingencies, as follows:

Project 1	Productivity improvement
Technical risk	New equipment failure
Contract risk	Supplier delivery failure
Performance risk	Poor estimates of productivity payoff

Project 2	Build Huntsville project
Technical risk	Construction design failure
Performance risk	Unforeseen labor issues
Cost risk	Inaccurate estimates of cost

TABLE 7.8 Huntsville Project 1 Risk Matrix

Risk	Impact	Probability (%)	Contingency	Cost of contingency
New equipment	Technical failure; schedule and cost	50 (unproven technology)	Modify equipment; new tooling	$40,000
Contract	Contractor failure; schedule and cost	75	Rebid contract; start over	$33,000
Productivity	Payoff not achieved; schedule	75	Reduced sales	$540,000
Total risk score				$613,000

TABLE 7.9 Huntsville Project 2 Risk Matrix

Risk	Impact	Probability (%)	Contingency	Cost of contingency
Construction design failure	Bad construction design, schedule and cost	50 (unproven technology)	Modify equipment; new tooling	$40,000
Bad space layout	Cost	75	Rebid contract; start over	$3,000
Low labor productivity	Productivity payoff not achieved	75	Reduced sales	$40,000
Inaccurate estimates of cost	Budget overrun	75	Reduce net income	$200,000
Total risk score				$283,000

Comparing the two risk scores, the Huntsville Project is the lower cost exposure due to risk. Therefore the risk analysis supports the choice of Project 2.

Huntsville: Risk-Based Scheduling

A risk-based schedule can be put together using the MS Project toolbar, the outcome is as shown in Table 7.10.

Risk and the Huntsville Project: Journey in Risk

In sum, the Huntsville Project is instructive because there are so many risks associated with the whole process, from business strategy through project review, based on earned value. Risk and benefit were integral to the project life cycle.

For instance, there were substantial risks associated with:

- The development of business strategies
- The selection of a project from the two candidates
- Project-level risk—overall risk associated with the two candidates
- The development of the task list
- The estimation of task duration
- The determination of resource requirements and assignments

TABLE 7.10 Huntsville Risk-Based Schedule Data

High-risk task	Expected duration	Optimistic duration	Pessimistic duration	Caclulated, risk-based duration
Select architect	55 days	20 days	80 days	65.83 days
Site selection	40 days	23 days	70 days	57.17 days
Truck fleet	66 days	50 days	89 days	78.67 days

- The linkages in the project schedule
- The performance of the key project team members
- The conflicts inherent in the project team
- The decision to use actual hours as an indicator of work accomplished
- The entry of data into MS Project
- Risks associated with each task

At each such step in the process, the risks and benefits of the situation at hand had to be measured and evaluated against the strategic objectives of the company. Some risks were documented (e.g., the risk matrices in the solution), but in other cases the risks were considered, ranked, and decisions made without analysis. This is the *real world* of risk since a thorough analysis of every risk would not be feasible.

Chapter

8

Customer-Driven Project Management, TQM, and Risk

The purpose of this chapter is to illustrate the importance of *seeing* risk from the customer's perspective and recognizing that risk is inherently a customer and quality issue as described in total quality management (TQM) concepts. As the major stakeholder and project sponsor, the customer/client *pays the price* at the end of a project if risk is not well managed. In the context of the three legs of the total quality management stool, risk is associated with: (1) customer expectations and satisfaction, (2) empowering the project team, and (3) the effectiveness of continuous process improvement.

Customer-Driven Risk Management

A customer-driven project team is a team that responds to customers and manages customer satisfaction as a regular team function. Risk management is an overall obligation of the customer-driven project team. The teams continually assess risk at the project level and in each task of the project—not only in terms of time and cost, but also in terms of the technical feasibility. Again, the customer-driven lead team establishes the system for risk management. This risk management system influences the use of the other project management tools and techniques.

No project is without risk. That is, the probability that a given process, task, subtask, work package, or level of effort cannot be accomplished as planned. Risk is not a question of time; it is often a question of feasibility. It pays in the development of the WBS to assign risk factors to each element of the WBS, separate from the assignments of schedules and milestones for the work. Risk factors can be assigned based on uncertainty, technological feasibility, availability of resources, or competition. Later, the elements with high-risk factors are given close attention by the project manager, whether or not they are on the critical path itself. Special attention is given to customers, users, and clients.

One of the most effective ways to deal with risk is to develop customer-driven contingency plans, or parallel courses of action—and changes in the WBS—that come into play if the task cannot be accomplished as the customer sees it. Contingency planning is based on "satisficing," as Herbert Simon called it, rather than optimizing, and recognizes that in developmental projects, some things turn out to be impossible dreams.

Customer-driven contingency plans are carried out by the project manager, based on the level of risk assigned to a particular element of the WBS. When testing a new technique or approach, for example, a manager might estimate a 15 percent chance of successfully completing Task 1 and satisfying the user, because few new techniques work well the first time they are attempted. Similarly, there may be only a 10 percent chance of successfully completing Task 2 because of the organization's previous experience with a similar problem. Hence, there is only a 1.5 percent probability of even reaching Task 3 if its performance depends upon successful completion of the preceding tasks. Once Task 3 is completed, however, it may be implemented again and again, depending on how many problems have to be solved.

Illustration of Risk Management—the Defense Risk Program

The U.S. Department of Defense (DoD) has for many years provided standard templates to contractors for reducing risk in system development. The DoD uses templates directed at the identification and establishment of critical engineering processes and their control methods. For each of the critical engineering processes a critical path template is provided. The template addresses the following areas for each critical engineering process:

- Area of risk
- Outline for reducing risk
- Timeline

TQM Template. TQM is defined as an organized process of continuous improvement by private defense contractors and DoD activities aimed at developing, producing, and deploying superior material and products. The primary risk in reaching and sustaining this superiority is failure to manage with a purpose of constantly increasing intrinsic quality, economic value, and military worth of defense systems and equipments. Note the focus on management first, not technical risk. The armed forces and defense industrial entities may not attain a lasting competitive military posture and long-term competitive business stature without a TQM at the highest levels. TQM is applicable to all functions concerned with the acquisition of defense material, supplies, facilities, and services. Being satisfied with suboptimum, short-term goals and objectives has adverse impacts on cost, schedule, and force effectiveness.

A short-term approach leads to deteriora-tion in the efficacy of specific products, the firms that produce them, and the industrial base overall. A major risk is also entailed with the inability to grasp and respond to the overriding importance attached to quality by the "customer" or user activities.

DoD outline for reducing risk

- The organization has a "corporate level" policy statement attaching highest priority to the principles of TQM. This policy statement defines TQM in terms relevant to the individual enterprise or activity and its products or outputs.
- The corporate policy statement is supported by a TQM implementation plan that sets enduring and long-range objectives, list, criteria for applying TQM to new and ongoing programs, provides direction and guidance, and assigns responsibilities. Every employee at each level plays a functional role in implementing the plan.
- All personnel are given training in TQM principles, practices, tools, and techniques. Importance is placed on self- initiated TQM effort.
- The TQM effort begun in the conceptual phase of the acquisition cycle is vitally concerned with establishing a rapport between the producer and the user or customer and a recognition of the latter's stated performance requirements, mission profiles, system characteristics, and environmental factors. Those statements are translated into measurable design, manufacturing, and support parameters that are verified during demonstration and validation. Early TQM activity is outlined in the Design Reference Mission Profile template and Design Requirements template. The Trade Studies template is used to identify potential characteristics, which would accelerate design maturity while making the design more compatible with and less sensitive to variations in manufacturing and operational conditions.
- Design phase TQM activity is described in the Design Process template. Key features enumerated include: design integration of life cycle factors concerned with production, operation, and support; availability of needed manufacturing technology; proof of manufacturing process; formation of design and design review teams with various functional area representation; and use of producibility engineering and planning to arrive at and transition a producible design to the shop floor without degradation in quality and performance. The Design Analysis template and Design Reviews template provide guidance in identifying and reducing the risk entailed in controlling critical design characteristics. Both hardware and software are emphasized (refer to the Software Design template and Software Test template). A high-quality design includes features to enhance conducting necessary test and inspection functions (refer to the Design for Testing template).
- An integrated test plan of contractor development, qualification, and production acceptance testing and a test and evaluation master plan (TEMP)

covering government-related testing are essential to TQM. The plans detail sufficient testing to prove conclusively the design, its operational suitability, and its potential for required growth and future utility. Test planning also makes efficient use of test articles, test facilities, and other resources. Failure reporting, field feedback, and problem disposition are vital mechanisms to obtaining a quality product.

- Manufacturing planning bears the same relationship to production success as test planning bears to a successful test program (refer to the Manufacturing Plan template). The overall acquisition strategy includes a manufacturing strategy and a transition plan covering all production-related activities. Equal care and emphasis is placed on proof of manufacture as well as on proving the design itself. The Quality Manufacturing template highlights production planning, tooling, manufacturing methods, facilities, equipment, and personnel. Extreme importance is attached to subcontractor and vendor selection and qualification including flow down in the use of TQM principles. Special test equipment, computer-aided manufacturing, and other advanced equipments and statistics-based methods are used to quality and control the manufacturing process.

Timeline. TQM is used throughout the product life cycle. TQM-oriented defense contractors and government activities concentrate on designing and building quality into their products at the outset. Successful activities are not content with the status quo or acceptable level of quality approach. These activities respond to problems affecting product quality by changing the design and/or the process, not by increasing inspection levels. Reduction in variability of the detail design and the manufacturing process is a central concept of TQM and is beneficial to lower cost as well as higher quality. Defect prevention is viewed as key to defect control. Astute TQM activities are constantly on the alert to identify and exploit new and proven managerial, engineering, and manufacturing disciplines and associated techniques.

DoD manuals stress the following TQM principles:

- Total commitment to quality
- Continuous improvement
- Involvement of many functions
- Long-term improvement effort
- Customer focus

TQM principles include company action that:

- Produces a policy statement (vision/mission)
- Pursues a TQM environment
- Stresses a TQM implementation plan

- Fosters ownership
- Advocates training
- Includes quality as an element of design
- Encourages measurements
- Includes everything and everyone
- Nurtures supplier and customer relationships
- Encourages cooperation and teamwork

Portfolio and Program Management

DoD policy first articulated the importance of project selection and portfolio management in the 1980s and has refined the concept. The initial and most disruptive risk involved in program management lies in the inability to choose the right portfolio of projects in the first place. Fit and consistency with strategic plans, cash flow analysis, rate of return, company competency (Can we make this product?), lack of business analysis on markets, technical feasibility, and legal risk are key factors contributing to successful risk management.

- *Fit with company strategy.* Alignment with strategy is a key risk challenge because projects that might otherwise be attractive might not be part of the company's plans for growth and competence. The weighted scoring model is one tool for assuring "fit with strategy," but the risk here is in the inability to measure whether, in fact, a candidate project is going to help implement a company strategic goal.
- *Cash flow.* Since profitability and rate of return is key to company growth, a project must be planned out over the long term to ensure net income growth. This means that candidate projects must be "fleshed out." The risk here is that cash flows are misestimated and that key decisions and assumptions behind the decision are not made explicit.
- *Consistency with company competency.* If the company does not have experience in a given area, it does not matter whether the project is consistent with strategy and cash flows are promising. Company capacity to perform is directly related to past experience and capacity—"sticking to the knitting" is still a major principle of success and therefore a useful risk management concept.
- *Market analysis.* If the market and future demand for a given project outcome or product/service are not well researched, the risk is that an efficient project may meet schedule, cost, and quality objectives, but produce a product which is not marketable. Therefore the focus in risk planning must be in the adequacy of early business market analysis and research.
- *Technical feasibility.* Since technical feasibility is key to new systems, if a project involves new, unproven technology there is inherent risk in the project.

Technical feasibility can be offset by embedded risk measures as described earlier in the product development process.

- *Legal risk.* Legal and regulatory risk is attributable to changing government regulations and legal issues involved in liabilities for a given product. The risk here is that the company fails to understand and anticipate legal and regulatory constraints on a product or service.

Value of Customer-Driven Risk Management

Customer-driven risk management captures the critical importance of seeing risk in the eyes of the customer and client. A focus on customer risk can uncover uncertainties and risks in a project that are not apparent in an internally focused risk management process. For instance, in developing a software product, the customer focus may be grounded in compatibility or interface of systems, while the internal, project-oriented focus for software development might well be in conformance with performance requirements without much attention to interface issues.

Risks in customer expectation, needs, and requirements

Program and projects face customer risk is three areas and each challenges the risk management process:

- Customer expectations
- Customer needs
- Customer requirements

Customer expectations. Sometimes customers expect more than they specify in written specification documents. Expectations change from new information uncovered in the project itself. The risk associated with customer expectations reflects the inherent value of projects themselves; as products and service outcomes are produced and information is made available to the customer that uncovers new opportunity, customers sometimes change their minds.

Customer needs. What the customer needs is not necessarily what the customer expects or requires. Needs suggests analytic data on customer needs, an objective view of needs underlies the project process, but needs are seldom differentiated from expectations and requirements.

Customer requirements. Requirements are the actual specifications for the product or service outcome. Sometimes requirements are drawn up by project teams based on what is feasible rather that what is required. The risk here is that the requirements document does not adequately capture customer need.

Risks in customer and sponsor relationships

There are two basic risks inherent in the relationship with project customers and sponsors. They are:

1. Losing the commitment and support of sponsors. Project sponsors can withdraw support from a project for a variety of reasons unrelated to the project itself, e.g., financial issues, market turns, changes in organizational leadership, mergers and acquisitions. The risk is that the project manager will be surprised by such a withdrawal of support and have no contingency or "workaround" action to offset it. Again, it is the ability of the project manager to foresee factors that might lead to loss of sponsor support that will lead to successful management of this risk.
2. Managing sponsor resistance to change. Sometimes a new product or service will produce changes in a customer or sponsor organization that can be dysfunctional. Thus the risk is that the project manager again is surprised by resistance in a sponsor or customer organization to the very product or service the customer specified.

In both of these cases the lesson is this. Build and maintain a close relationship with your customer and sponsor to help anticipate risks and issues *for the customer* before the customer sees them.

Chapter

9

Strategic Planning and Risk—The Eastern Case

As indicated earlier, risk is inherent in the nature of the business itself. Business planning aimed at developing a business strategy considers various risks and threats to its success in the planning process. This chapter uses the case approach by addressing how a real company, termed the Eastern Company for purposes of this case, handles risk in its business planning process. Eastern is a global manufacturer and distributor of aluminum products.

Typically, the Eastern Company faces major competition and challenge from a global aluminum market and from foreign manufacturers who regularly "dump" aluminum into western markets at very low prices. Thus there is continuous risk in the business from forces out of the control of internal company and project management. To address the risks inherent in its business, Eastern prepared a risk-based strategic plan.

Eastern faces eight risks and has developed eight strategic goals to address them.

Risk 1. Required electric power will not be available at an affordable price.

Strategy 1. Secure economically priced power to reduce the risk of power shortage.

Risk 2. Cost increase in aluminum manufacturing will increase faster than margin.

Strategy 2. Secure other resources at reasonable costs to offset the risk of cost escalation.

Risk 3. Customers will not be satisfied with Eastern's products.

Strategy 3. Cultivate customer awareness and promote customer satisfaction to avoid customer satisfaction risk.

Risk 4. Eastern's working environment will prove to be unsafe and the company will experience substantial loss of workforce and finance as a result.

Strategy 4. Create a safe working environment to control the risk of worker injury and associated costs.

Risk 5. The Eastern workforce will not grow with the technology available for continuous improvement.

Strategy 5. Build a responsible and knowledgeable workforce to avoid the risk of workforce instability.

Risk 6. Eastern will not act to improve the technology of manufacturing in time to keep ahead of competitors.

Strategy 6. Improve technology and plant equipment to produce products more efficiently to control productivity risk.

Risk 7. Pollution from Eastern facilities will lead to noncompliance with government environmental requirements.

Strategy 7. Improve Eastern's impact on the environment to avoid the cost of pollution and noncompliance.

Risk 8. Increasing waste in the manufacturing process and workforce will lead to uncontrolled costs.

Strategy 8. Reduce waste and non-value-added costs to control the risk of wasted effort.

Eastern recognized the need to directly take action to sustain its ability to successfully compete on a continual basis in the world aluminum marketplace. The assumption was that despite the fact that Eastern employees, in general, were dedicated to providing the highest-quality products and services to the customer at a competitive price, and to providing a positive return for their owners' investment, they were heavily unionized. The company committed to the principle that: *We will not be able to step up to those challenges unless our employees—and the union—can see where we are going and why, and have the opportunity to "buy-in."* It is through this strategic plan and its communication plan that they saw that they could accomplish alignment and reduction of their considerable risk exposure.

The strategic plan was communicated continuously throughout the plant through special meetings and focus groups to ensure that all employees understood it and could relate it to their work. Employees were encouraged to document actions they or their teams were taking to accomplish or support particular initiatives. This process would continue as we updated the plan annually and realigned our policies, procedures, and organizational structure to accomplish the plan.

This strategic plan was developed by the directors of Eastern with support from area managers.

Commitment and Partnership

Eastern management and the union stated directly that they were committed to this mission for the organization. It was clearly recognized that by working together to accomplish this mission, the interests of all participants would be served. All management and employees would benefit from long-term job security, job enrichment, and the monetary rewards that result from a successful business, which was able to manage its risks. Eastern's stakeholders and owners would benefit from the product recognition and profitability gained by producing superior goods and services. Eastern's customers would benefit from the high quality and service levels delivered to them. Finally, the community was to enjoy a stable revenue base from the success of Eastern and from the skills and services individual employees offered.

Stakeholder relations

The company stated that Eastern's *stakeholders* were people, organizations, or groups of people who have a vested interest in the success of the company. Their major stakeholders were:

- Employees, who seek continued employment and income, quality of work life, and opportunities to learn and develop—their perceived risk was related to job security and lack of growth and development and marketability
- Customers, who seek quality products at low cost and reliable delivery—their perceived risk was related to product price and quality and timing, but mostly price. Cost was a major issue as competitors dumped quality aluminum at lower prices
- Owners, who seek return on investment and continued viability—their risk was grounded in stock value
- Regulators, who seek compliance with laws and regulations—their risk was in noncompliance with regulations and the cost of enforcement and litigation
- Community, who seek contributions through taxes and service, and minimal environmental impact—their risk exposure was in losing the industry tax base but having to pay pollution and environmental control costs
- Suppliers, who seek to meet Eastern requirements and continue business with Eastern—their perceived risk lies in their inability to meet Eastern contract requirements and having to share more of the risk in contracted work than they can handle

To illustrate the documentation of a risk-based strategy, the following document contains an executive summary, situation analysis, and a detailed description of eight key strategies. The situation analysis provides a framework for the strategies including mission and goals, management direction, SWOT analysis, and linkage to the parent company strategic plan. The eight strategies are supported by specific initiatives and a system to measure achievements of those initiatives.

This strategic plan for Eastern Aluminum Company covered a 5-year period, from 1996 to 2000, and thus will help to guide the company and its employees into the twenty-first century. As the general long-term pathway to growth and profitability, the plan presents the company's approach to achieving Eastern's central strategic goal—to compete successfully on a continuing basis in the world aluminum marketplace. The plan served a wide variety of purposes, including support to ownership decisions; support to budgeting and resource allocation; guidance for management and employee planning, training, and education; and support to long-term capital investment planning. A major element of the strategic planning process that produced this document is the communication of the plan and its underlying vision, assumptions, and values to our employees.

The plan explored Eastern's current strengths, weaknesses, opportunities, and threats, and presented and discussed eight basic key strategies, initiatives, and measures to accomplish the central strategic goal.

Eight Strategies

Within the overall framework of the basic strategic goal to compete successfully in the world aluminum marketplace, and consistent with the parent company's strategic objectives, eight key strategies were at the heart of this strategic plan:

1. *Secure economically priced power.*

   Eastern would find ways to lower its power costs through a variety of strategies, including building stronger partnerships with power companies and state and local governments, and through exploration of independent options for generating less expensive power.

2. *Secure other resources at reasonable costs.*

   As the cost of materials rises, Eastern planned to find low-cost sources for raw materials as well as explore approaches for using lower-grade materials. Eastern would take the initiative to ensure that effective partnerships are built with quality suppliers.

3. *Cultivate customer awareness and promote customer satisfaction.*

   Eastern would work to educate employees about customers and their requirements and promote closer ties with customers. Greater appreciation of customers would give employees more incentive for addressing future customer requirements and connecting their daily work more clearly with the "value chain" to the customer.

4. *Create a safe working environment.*

   Eastern was working to improve its safety record through enforcement of safety and health rules and regulations. Employees would be better educated and trained to understand safety implications of their work. Safety compliance would be considered a major performance standard for all employees.

5. *Build a responsible and knowledgeable workforce.*

   Facing a major workforce turnover in the next 5 years, Eastern placed special emphasis on strategies to build a more responsible and skilled workforce, to improve the partnership with the union, to improve performance and productivity, to lower labor costs, and to find better ways to work together through teamwork. They recognized that if this strategy—grounded in the commitment to building a team-based organization—is not accomplished, Eastern could not thrive and grow even if the other strategies were accomplished.

6. *Improve technology and plant equipment to produce products more efficiently.*

   Eastern prided itself on its leadership in technology and technical innovation and plans to continue this industry leadership. Eastern was managing several capital improvement projects to make major breakthroughs in productivity and quality. Eastern felt it was demonstrating to its customers and its employees through these improvements that major investments are being made in the plant to meet the challenges of the future global marketplace.

7. *Improve Eastern's impact on the environment.*

   Through strict compliance with federal, state, and local environmental standards, Eastern would continue to respond to and anticipate environmental impacts and address them. Special emphasis was made to meet new clean-air requirements.

8. *Reduce waste and non-value-added costs.*

   Eastern continued to pursue quality and process improvement initiatives to eliminate unnecessary costs due to accidents, rework and scrap, outdated positions and job requirements, and equipment damage. Employees would continue to be trained and educated in process improvement and reengineering to streamline the way work is accomplished.

Overview

Eastern had already been turned around from a high-cost swing plant, with a confrontational labor atmosphere, to a much more competitive operation practicing effective and efficient management and supervision, worker empowerment, and self-directed team concepts. However, there was new urgency to ensure that all employees understood that the plant would grow only "by permission" from future customers, and only if it continuously improved its productivity, quality, and internal cohesion and teamwork across departments. The following text discusses how they were positioned to compete in the future.

Strengths, Weaknesses, Opportunities, and Threats

The following discussion covers Eastern's strengths, weaknesses, opportunities, and threats.

Strengths

Eastern had made a concentrated effort to retain its competitive position in the marketplace through technology. Its major strength is its ability to produce quality products continuously, focus on technology and capital improvement, and keep wages and salaries relatively high for its employees while controlling costs. Capital improvements and improved management and team practices have made it possible to achieve record premium production in the recent past. Eastern continued to demonstrate its leadership in technological improvements and plant capital investment.

Eastern had experienced the longest run at full capacity in its history, remaining one of the few North American smelters not curtailed due to the recent metal surplus caused by the flow of aluminum into the world market from the Commonwealth of Independent States. Eastern continued to show resilience and responsiveness in the face of changing market conditions.

Eastern was working hard to empower our workforce—improve their knowledge, skills, and responsiveness. They sought to align their incentive and reward programs, partnership practices with hourly employees, performance appraisal systems, and quality and process improvement initiatives with our key long-term strategies. They faced the future turnover of the workforce with a strong commitment to use the opportunities that changes bring to build a leaner, more integrated, and productive plant team.

At the heart of its strength was Eastern's traditional core competency to choose, operate, and improve process technology effectively; to produce a variety of difficult-to-produce premium products; and to understand and meet customer needs. Whatever initiatives Eastern undertook, it knew it had to continuously improve these drivers of success.

Weaknesses

Eastern's products (primary, slab, billet, tee, and foundry pig) were priced by the worldwide commodities market. High quality and excellent service of these products would ensure a positive customer relationship, but Eastern could not control the selling price of the finished product.

Eastern knew that it was a high-cost plant compared to other producers, primarily because of the age of the facility and technology, and because of high wages, salaries, and fringe benefit levels. Because Eastern had little or no control over the market price, the cost of producing aluminum became a key determining factor in remaining globally competitive. In fact, 75 percent of all aluminum in the world was being produced at a cost lower than Eastern.

Eastern faced major challenges in turning over its workforce and creating a more energetic and knowledgeable workforce team. Past practices had not always inspired employees to align themselves with the plant's best interests and commit themselves to continuous improvement through teams.

Eastern needed to improve its ability to learn and document its successes, in short, to become a "learning" organization. Past practice had not always taken advantage of what the organization had already learned through the years.

Opportunities

Demand for aluminum was continuing to rise; supplies of aluminum had increased each year with primary aluminum products now sold on a worldwide basis.

Eastern had the opportunity to position itself at the midpoint on the world cost scale, the point at which 50 percent of world production costs would be higher than Eastern's. In achieving this position, Eastern could take more advantage of its high-quality products and services and its improving productivity.

A reduction of 4 cents per pound by 1999 would have placed Eastern in that competitive position, keeping mind that other aluminum plants are also attempting to reduce their costs.

Eastern's major opportunity was to improve its process efficiency and productivity through a combination of technology and capital improvement, building a more efficient workforce and reducing labor costs. The 4 cents per pound cost reduction could be achieved by:

1. Conversion of potlines (production lines) to a new "point feed" technology already underway.
2. Reduction of man-hours per ton by 15 percent from 1996 to 2000.
3. Reduction of non-value-added costs wherever possible through process improvements, total quality management, ISO 9000 certification, and other quality initiatives.

Eastern had a major opportunity to improve its human resource practices and programs as the plant transitions its workforce in the coming five years, both through better training and development of supervisory and hourly employees, and better, more effective assessment and hiring practices.

Threats and Risks

If Eastern did not continually reduce costs, their position would worsen because:

1. New plants with lower costs would open.
2. Existing competitive plants would reduce cost and improve their cost position.
3. Other plants with higher costs would close, worsening Eastern's position.

The most critical of these risks was the possibility that power costs would continue to rise beyond Eastern's capacity to absorb them. This scenario represented the most significant threat to Eastern's continued growth and had to be avoided. In addition to power costs, the long-term cost of coal could be another important threat to Eastern growth, as well as unanticipated environmental regulations, particularly from the federal clean air act.

In addition, although Eastern had made major progress in building a more team-based culture, the process could not be slowed by resistance to change and failure to be clear about new roles and functions. Therefore, one source of threat and risk was clearly from within, the threat of slow deterioration of the momentum of teamwork and process improvement already underway. Such a step backward could always happen as a result of neglect and a lack of trust and respect in the organization.

Eastern's Strategic Plan

Eastern's strategic plan was an integrated set of strategies, initiatives, and measures supporting an overall goal of competitiveness. Figure 9.1 presents a graphic depiction of the company's eight key strategies. Each strategy was seen as serving the central goal of world competitiveness, but each strategy was also inextricably tied to the others, indicating a strong interdependency of all plan elements. If any one strategy and risk-reduction plan was not accomplished, overall achievement of the goal suffered.

The plan describes plant strategies, initiatives, and measures of success. Initiatives are programs and projects now underway or planned to help accomplish a particular strategy. Measures are indicators of progress and will be used to monitor the achievements of the eight key strategies.

Underlying elements of the risk-based strategic plan

Five major elements formed the basis for this risk-based strategic plan—mission, commitment and partnership, driving forces, core competencies, and stakeholder relations. They are discussed below.

Mission. Eastern's mission was to be the most cost-effective producer of the highest-quality primary aluminum products shipped on time to its customers,

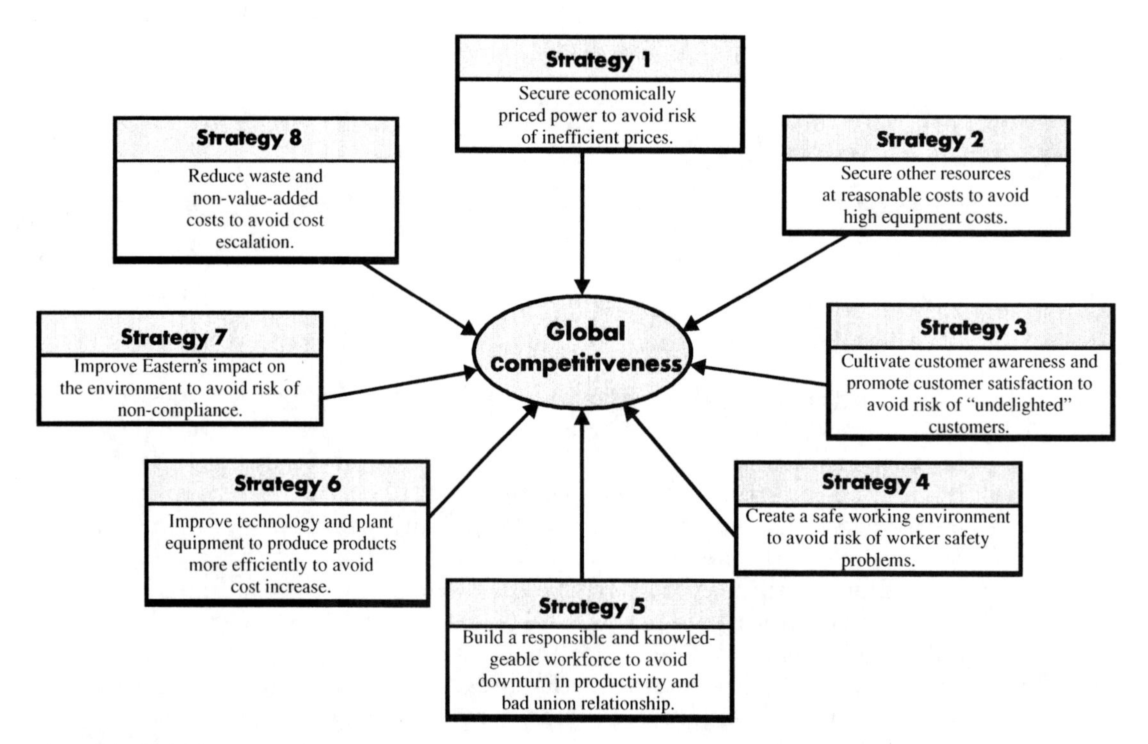

Figure 9.1 Eight key strategies.

with optimum utilization of resources. They placed special emphasis on their employees and their role in defining the company's mission and on good community relations. They recognized that accomplishing the mission involved a never-ending journey of continuous improvement.

Commitment and partnership. Eastern management and the union indicated that they were committed to this mission for the organization. It was clearly recognized that by working together to accomplish this mission, the interests of all participants were best served. All management and employees would benefit from long-term job security, job enrichment, and the monetary rewards that resulted from a successful business. Eastern's stakeholders and owners would benefit from the product recognition and profitability gained by producing superior goods and services. Eastern's customers would benefit from the high quality and service levels delivered to them. Finally, the community would enjoy a stable revenue base from the success of Eastern and from the skills and services individual employees can offer.

Driving force: production capability. An underlying element in this strategic plan was the single most important driver of company success—its capability to convert resources effectively into products through highly organized and managed production processes. Their value-added for the future will continue to be their capacity to produce products continuously.

Core competencies and risk contingencies. Three core competencies separated Eastern from its competitors:

1. *Its capacity to effectively choose, operate, and improve process technology.* Its ability to keep up with technology change was rooted in its ability to anticipate technology risk stemming from out-of-date technology.
2. *Its capacity to produce a variety of difficult premium products.* Its ability to change its production systems quickly was rooted in its ability to anticipate the risks of change in product requirements and plan for them.
3. *Its capacity to understand and service customer needs.* Its capacity to understand its customers and especially to anticipate and manage customer services to reduce the risk of failed customer service expectations.

Eastern would strive to maintain and build on these core competencies.

More on eight key strategies

Eastern identified eight key strategies to carry out its central strategic goal of global competitiveness. Each strategy was carried out through several initiatives and was monitored by the measures presented and discussed below.

Strategy 1—Secure economically priced power. The cost of power was a major factor in Eastern strategic plan. In its partnership with the community, power

TABLE 9.1 Measures and Risks: Cost of Power

Initiatives	Measures	Risks
Address power pricing issues by: Maintaining relationships with Potomac Edison, Public Service Commission, People's Council, local and state government Investigating power-wheeling sources and benefits Developing alternative power sources, including self-generation Increasing community support for reducing Eastern's power costs Eliminating power modulation	Reduced power costs by 2 to 4 mills per kilowatt hour (approximately $6 to 12 million/year) Favorable public response and concern for Eastern's power pricing issues	That relationships with stakeholder agencies would deteriorate That power-wheeling sources (independent sources of power created by deregulation) would not provide lower prices That self-generation of power would fail either from technology problems or cost That the community would not support Eastern That power modulation—the practice of energy providers to reduce power could not be anticipated

companies, and state and local government, Eastern would develop support for its efforts to continue to compete in the world aluminum marketplace. Eastern planned to negotiate lower power costs and to explore independent options for generating less expensive power (Table 9.1).

Power costs became the key factor in maintaining competitiveness because of impending major increases in costs from new sources. Eastern management was working closely with utilities, government officials, the community, and other power sources to ensure that it can achieve independence in power generation should that be necessary. Power-wheeling sources and benefits were being pursued as well as self-generation options.

While this issue is beyond the scope of any one employee, special attention was given to communicating the power cost issue to all employees so that they could understand the urgency of the situation and help Eastern to achieve power independence.

Strategy 2—Secure other resources at reasonable costs. The cost of materials continued to rise and had the potential to erase savings created by increased productivity and reduced power costs. Eastern planned to find low-cost sources for raw materials as well as explore approaches for using lower-grade materials. Eastern would continue to manage human resources costs through employee attrition and retirements (Table 9.2).

Key raw materials (alumina, aluminum fluoride, petroleum coke, and liquid pitch) were purchased for all parent company smelters by the same parent office. These costs were rising to a point such that the Eastern's overall cost effectiveness was threatened.

This issue challenged the company's capacity to find and use lower-grade materials such as calciner fines and lower grades of petroleum coke. The company would continue to acquire both raw materials and supplies from the most efficient sources while assuring quality. This involved forming partnerships

TABLE 9.2 Resources: Measures and Costs

Initiatives	Measures	Risks
Obtain raw materials such as petroleum coke, pitches, alumina, and hardeners Secure high-quality supplies from the most economical sources	Maintain or decrease current raw material costs	That raw materials would not be available on a just-in-time basis
Manage human resource (labor) costs through attrition and retirements	Reduce man-hours per ton by 15 percent by the year 2000 Contribute to overall efficiency and productivity	That human resource costs would inflate and attrition goals would not be achieved
Explore innovative approaches to using lower-grade materials such as calciner fines and lower grades of petroleum coke	Maintain or decrease current raw material costs	That lower-grade materials would be acquired

with suppliers to limit the number of such sources. This would accomplish two objectives—holding down costs and minimizing purchasing and warehousing requirements.

As a major cost element, labor costs had to be controlled while productivity was enhanced through capital improvements and better management, team, and individual performance. Reduction of man-hours per ton by 15 percent by the year 2000 was a major measure of success in reducing risk exposure.

Strategy 3—Cultivate customer awareness and promote customer satisfaction. Eastern continued to provide consistent and high-quality products and services to end-users and customers. The company would work to ensure that all employees were aware of customers and their needs. Emphasis on the customer would encourage the development of new products and services and help Eastern establish a larger market niche. Eastern would look to external stakeholders to verify gains made in employee customer awareness and customer satisfaction (Table 9.3).

Eastern had to establish a market niche in high-quality, premium products to remain a viable company and successfully compete. To meet this demand, Eastern had to work closely with its ownership to identify future customer needs. Eastern would continue to work with parent company marketing teams in the areas of initial order processing, customer team visits, and customer surveys.

Eastern would also make it easier for customers to deal with the plant. Increased use of bar code systems and electronic data interchange would be planned, establishing a "seamless" electronic relationship with prospective customers. More attention would be paid to promoting our laboratory and metallography capabilities.

The continuing move to quality worldwide was having its impact on the company. More customer inquiries, such as from the automotive industry, were

TABLE 9.3 Employee Awareness: Measures and Risks

Initiatives	Measures	Risks
Enhance employees awareness about customer and end-product satisfaction	Third-party assessment of employees' customer awareness Recognition through accreditation and quality audits (American Association for Laboratory Accreditation, and the like)	That employees were not able to connect their success with company success in end-product quality
Selectively diversify products and services to support market expansion	Capacity to change products Number of customer assists through the Metal Quality Group	That its product mix could not be diversified
Support parent company marketing strategy Market services to make customers aware of Eastern's capabilities	Inventory Management System (AIMS) data Customer team visit comments Customer satisfaction data	That the parent company strategy was not consistent with Eastern's strategic plan and core competence
Focus on individual customer demands in metallurgy, product chemistry, packaging and delivery requirements through process improvement Develop a long-term cast house plan and monitoring systems	International Standards Organization (ISO) 9000 registration QS 9002 accreditation Monitor customer claims and contacts about technology services and products Review customer satisfaction survey results Improved product turnaround indicators	That Eastern's process improvement efforts were not successful because of personnel and union disincentives
Set up cross-functional teams to increase awareness of internal customers	Internal customer satisfaction surveys Extent to which internal customer requirements are met	That cross-functional teams would not work because of internal conflicts and role definitions

expected regarding our quality standards. This development prompted efforts to maintain registration and refine documentation to both ISO 9000 and American Association of Laboratory Accreditation, and for attaining QS 9000 and 14000 certification as well. Increased cycle time was becoming a major customer expectation, generating internal plans to develop systems to measure order entry, production scheduling, and shipping performance.

Finally, because many employees did not have a direct relationship with customers and customer needs, the company was undertaking a program to enhance employee appreciation of customer needs. This program included use of a third-party organization to monitor employees' understanding of these issues.

Strategy 4—Create a safe working environment. Eastern had significantly reduced accidents in recent years and needed the support of employees and management to continue these safety efforts. In addition to developing and implementing state-of-the-art safety procedures and guidelines, the company needed to enforce safety and health rules and regulations *consistently* (Table 9.4).

Eastern recognized its responsibility and accountability for the safety and health of each employee and for the preservation of property and equipment.

TABLE 9.4 Safe Working Environment: Measures and Risks

Initiatives	Measures	Risks
Eliminate safety and health hazards by:	Decrease accident incident rate	That safety initiatives would not be accepted and implemented by employees and managers
Upgrading engineering standards, safety features and ergonomics consistently	Stay within accident and safety scorecard budget	That safety guidelines and regulations would shift substantially
Promoting employee awareness	Improve safety severity ratio index	
Updating Joint Safety and Health Committee guidelines	Rate of completion of items on safety list and audits	
Enforcing rules and regulations	Improvements in efficiency and job performance	
Increasing team and employee accountability for safety and health	Improved plant safety performance record	
	Increased safety gain sharing payout	

The company would continue to incorporate into the design and operation of all facilities safeguards and procedures that will minimize risks of personal injury and loss of property and equipment. Management was responsible and accountable for the safety and safe work conduct of all employees who report to them. Employees were equally responsible and accountable for safe practices as well as assisting in the ongoing safety program by reporting unsafe practices, procedures, or conditions when they were observed.

As indicated in the initiatives underway as part of this strategy, Eastern was giving special priority to upgrading engineering standards to reflect safety requirements and criteria. In some cases, this could have meant added cost and time constraints on planned capital projects, an expense well worth the investment in a safer working environment.

Strategy 5—Build a responsible and knowledgeable workforce. By increasing the skills and abilities of individuals, teams, and supervisors and empowering them, Eastern would be able to increase productivity, reduce operating costs, solve personnel problems, and increase teamwork across the entire plant. Initiatives in support of this priority included training and developmental opportunities in support of self-directed work teams (Table 9.5).

Strategy 5 held the key to successful achievement of the balance of the other strategies—the building of a workforce and organization that: (1) was aligned with the strategic direction of Eastern; (2) was structured, capable, and motivated to improve performance; and (3) worked together across departments to provide a "seamless" process of production and quality.

In building a flatter, more streamlined workforce, the company's strategy in the past had been to press for reduced manning and more teams and teamwork. As a result, many teams had been generated and trained to take responsibility to solve problems and make decisions necessary to keep their process operating at peak efficiency. Supervisory and hourly positions were reduced and roles and functions were changed.

TABLE 9.5 Workforce Teams: Measures and Risks

Initiatives	Measures	Risks
Develop or continue: Continuation of empowered, self-directed work team development (decision-making and responsibility) New performance appraisal system for salaried employees Development planning Knowledge and skills training for bargaining unit employees Supervisory development program Strategic plan communication process Conversion to parent salary structure New bargaining unit job classification (stemming from the labor contract) HR strategic plan	Better communication and coordination within team members and between supervisors and teams Innovative, timely, and sound employee and team decision-making Better use of tools, equipment, and raw materials Employees will be prepared to assume new responsibilities as a result of developmental exposure Enhanced partnership agreement Increase in ideas and solutions from employees Reduce man-hours per ton	That self-directed teams would not work in the unionized work setting That employees would not act on incentives to train and develop new skills That the strategic plan communication plan is not effective in improving employee support of company goals
Offer developmental opportunities to sustain employee education and growth through: Mentoring Inside training Outside technical managerial training Opportunities to manage	Successful development planning Track progress through training records Enhanced employee performance	That the plant could not implement mentoring and training initiatives because of the company culture

Now in the spirit of building the total Eastern organization, the company's strategic emphasis would go beyond reduced manning and generation of teams. The strategy would be focused on organizational effectiveness—building the whole organization through a stronger linkage and alignment between management, supervisors, and bargaining unit employees. The opportunity before company management was to build new supervisory roles and functions into our new team-based organization, requiring development of leadership skills, better business and productivity management and monitoring skills, and more support for technical supervision and cross-department process improvement. Support services such as human resources management were present to help to lead the effort. Organizational barriers to effective supervision would be identified and eliminated. Organizational and training initiatives were underway to help supervisors function as the guiding force for day-to-day operations.

Development tools would include business and productivity management, process improvement, facilitating and mentoring opportunities, inside training, outside development (technical and managerial), and management opportunities within the organization. To focus on incentives, Eastern would review its performance appraisal and gainsharing structure to ensure that they were aligned with this strategic plan, and making improvements when called for.

To ensure effective communication, quarterly plant communication meetings would continue and more information would be provided to employees "online," especially in the area of human resources.

Strategy 6—Improve technology and plant equipment to produce products more efficiently. The company was managing several capital improvement projects to upgrade the condition of equipment and work processes at the plant. The company needed to continue these improvements while also employing sound capital project management skills. Eastern would work to speed up completion of these capital projects and to keep them within budget and quality requirements (Table 9.6).

As evidenced in the partial list of capital improvements above, Eastern was heavily engaged in upgrading its technological and equipment base in order to maintain its leadership and core competency. The company was a front-runner in keeping pace with required capital improvements to aging plant infrastructure. Improvements are underway in the product production lines, carbon plant, cast house, substation, and laboratory, and in general plant functions such as emission and noise control, and information system management.

TABLE 9.6 Capital Improvement: Measures and Risks

Initiatives	Measures	Risks
Complete capital program and budgets each year Conduct major maintenance projects and overhauls	Completion of capital improvements, including Conversion of potlines to point feed technology Substation life extension Cast house continuous homogenizing furnace Rod shop anode cleaner Ladle shop ladle cleaner Bake oven rebuild Potline capacity expansion Rebuilt remelt furnace Developed stack filter systems for metal treatment Facilities expansion Completed stamper upgrades for billet and slab	That capital budgets would not be completed That maintenance projects are not completed for a variety of reasons
Conduct research to ensure that Eastern adapts or incorporates improved or emerging technologies	Completion of research and development (R&D) projects within budget	That necessary research on emerging technologies is not conducted
Develop a stronger capital project management system through training and other developmental assignments	Improved capacity to complete projects on time within budget and schedule	That the project control system is not made operational

The focus for this strategic plan was the completion of capital projects within budget, schedule, and technical requirements. This meant developing a stronger capital project management system and employing more effective project management practices.

Strategy 7—Improve Eastern's impact on the environment. The company would continue to monitor its impact on the local environment. These efforts would be directed toward reducing environmental degradation and pollution (Table 9.7).

This strategy addressed the company's environmental and community relations practices. Eastern would continue to stay ahead of environmental requirements through two basic approaches—(1) being proactive in assisting regulators at all levels in developing sound and cost-effective regulations that both implement environmental legislation and meet the needs of community and the business and (2) planning and implementing capital improvements and operating measures to comply with environmental requirements, attempting at the same time to ensure that such improvements also contribute to overall plant productivity.

Costs of compliance would increase as well in the administrative areas of record keeping, reporting, training, planning, and monitoring, and in acquisition of necessary monitoring equipment, creating the need to streamline these systems. Eastern would continue to develop the capacity to prevent pollution through technology improvements and through a multimedia approach that

TABLE 9.7 Regulatory Compliance: Measures and Risks

Initiatives	Measures	Risks
Comply with federal, state, and local environmental regulations by: Providing proactive assistance to regulators Educating employees about regulatory requirements Promptly reporting noncompliance and correcting any violations Filing Title V air permit application	Eliminate incidents of noncompliance Monitor response time for identifying and fixing violations	That new regulations that Eastern could not respond to would be enacted
Participate in voluntary activities on environment, safety and health issues, such as EPA Greenlights, reducing greenhouse gases and PFCs, and noise nuisance reduction	Eliminate environmental, safety, or health complaints about the plant or its operations	That voluntary efforts do not improve community relations
Encourage environmentally sound industrial and agricultural growth	Partnerships with state and local agencies	That local growth objectives and dynamics would change substantially
Continue farm production	Farm production and maintenance of safe environmental practices	That the company's efforts at farm production around the fringe of the plant were unsuccessful

addresses losses of material to air, storm water runoff and solid or liquid waste streams.

Strategy 8—Reduce waste and non-value-added costs. Eastern continued to experience waste and non-value-added costs, such as safety and property costs related to accidents, rework and scrap, and equipment damage. Process improvement and problem-solving teams would continue to focus on reducing these costs (Table 9.8).

This strategy was in concert with strategy 5 to build a knowledgeable and productive workforce. Both were required to improve overall productivity. This strategy was key to improving the overall productivity of Eastern by eliminating waste and unnecessary work, for example, by reducing the cost of poor quality through process improvement and ISO and QS 9000 and 14000 documentation.

The company's quality and process improvement efforts started on the production floor where its quality was built-in through consistent practices and extensive use of statistical process-control methods. Eastern was committed to being quality-driven, not cost-driven, thus the quickest route to elimination of waste and non-value-added costs was "doing it right the first time." They looked to this strategy to be a major factor in lowering our operating expenses by 4 cents per pound.

TABLE 9.8 Costs of Waste: Measures and Risks

Initiatives	Measures	Risks
Involve quality and risk reduction teams in identifying and resolving quality problems in key production processes	Amount of rework and scrap by department on a monthly basis Stay within approved budget guidelines for rework and scrap costs	That the company teams were not successful in resolving quality issues
Minimize equipment damage by educating employees, monitoring equipment use, and enforcing rules for properly using equipment	Review monthly maintenance to ensure departmental accountability for responsible equipment use Stay within approved budget guidelines for equipment expenses	That equipment damage rates continued
Eliminate duplication of effort in administrative processes Process improvement/reengineering Encourage employee use of best practice techniques	Benchmark other processes Monitor process costs	That administrative redundancy and increase costs of operation continued
Improve inventory management of supplies and equipment (includes maintenance, production, and raw material in-process)	Reduce inventory by at least 5 percent	That inventory management initiatives were not successful because of internal plant or supplier performance limitations
Minimize waste generation and increase recycling	Waste product reductions	That increasing rates of waste production would continue

The teams would continue to identify and resolve quality problems in key production processes—a new focus would be placed on administrative and support processes to ensure that they were under review in the context of process improvement as well.

Communicating Strategy and Risk

The company prepared a communication program to promote the company strategy and to explain the risks inherent in the business and the local plant setting. The structure of that plan has been provided below.

Key strategic goal. Improve Eastern's capability to compete on a continuing basis in the world aluminum market place.

Explanation of strategies

Strategy 1—Secure economically priced power. The cost of power is a major factor in Eastern's strategic plan. In its partnership with the community, power companies, and state and local government, Eastern will develop support for its efforts to continue to compete in the world aluminum marketplace. The company plans to negotiate lower power costs and to explore independent options for generating less expensive power (Table 9.9).

Strategy 2—Secure other resources at reasonable costs. The cost of materials continues to rise and has the potential to erase savings created by increased productivity and by reduced power costs. The company plans to find low-cost sources for raw materials as well as explore approaches for using lower-grade materials. Eastern will also continue to manage human resource costs through attrition and retirements (Table 9.10).

Strategy 3—Cultivate customer awareness and promote customer satisfaction. Eastern continues to provide consistent and high-quality products and services to end-users and customers. Eastern will work to ensure that all employees are

TABLE 9.9 Strategy 1: Secure Economically Priced Power

What we will do	How we will measure achievement
Maintain relationships with Public Service Commission, People's Council, local and state government	Reduced projected power costs by 2 to 4 mills per kilowatt hour (approximately $6 to 12 million/year)
Investigate power-wheeling sources and benefits	Report to support decision
Develop alternative power sources	Alternative power contracts
Increase community support for reducing Eastern's power costs	Obtain favorable public response and concern for Eastern's power pricing issues as indicated by favorable legislative and regulatory decisions
Eliminate power modulation	Effective date, 1/1/97

TABLE 9.10 Strategy 2: Secure Other Resources at Reasonable Costs

What we will do	How we will measure achievement
Obtain lower cost raw materials such as petroleum coke, pitch, alumina, and hardeners	Continuous reduction in raw material costs
Secure high-quality supplies from the most economical sources	Reduction in rework due to supply quality
Explore innovative approaches to using alternative graded materials, such as calciner fines and lower grades of petroleum coke	Cost reduction
Reduce human resource (labor) costs through attrition and retirements	Reduce man-hours per ton (MPT) by 15 percent by the year 2000

aware of customers and their needs. Emphasis on the customer may encourage the development of new products and services and help the company establish a market niche. The company will look to external stakeholders to verify gains made in employee customer awareness and customer satisfaction (Table 9.11).

Strategy 4—Create a safe working environment. Eastern has significantly reduced accidents in recent years, and needs the support of employees and management to continue these safety efforts. In addition to developing and implementing

TABLE 9.11 Strategy 3: Cultivate Customer Awareness and Promote Customer Satisfaction

What we will do	How we will measure achievement
Enhance employees' awareness about customer and end-product satisfaction through a customer education and visit program	Third-party ratings and assessments of employees' customer awareness Recognition through accreditation and quality audits (American Association for Laboratory Accreditation, and the like) ISO 9002, QS 9000
Selectively diversify products and services to support Alumax market expansion	Capacity to change products as measured by new product cycle times Number of customer assists through the Metal Quality Group
Support parent marketing strategy Market services to make customers aware of Eastern's capabilities	Alumax inventory management system (AIMS) data on the results of inventory management initiatives Customer team visit comments Customer satisfaction data
Focus on individual customer demands in metallurgy, product chemistry, packaging, and delivery requirements through process improvement Develop a long-term cast house plan and monitoring systems	International Standards Organization (ISO) 9002 registration QS 9000 accreditation Monitor customer claims and contacts about technology services and products Review customer satisfaction survey results Improved product turnaround indicators
Set up cross-functional teams to increase awareness of the needs of internal customers within plant operations	Internal customer satisfaction surveys Extent to which internal customer requirements are met

TABLE 9.12 Strategy 4: Create a Safe Working Environment

What we will do	How we will measure achievement
Upgrade engineering standards, safety features, and ergonomics	Achieve 1 million accident free hours
Promote employee awareness	Stay within accident and safety scorecard budget
Update Joint Safety and Health Committee guidelines	Rate of completion of items on safety list and audits
Enforce rules and regulations	Improve efficiency and job performance
Increase team and employee accountability for safety and health	Improved overall plant safety performance record
	Increased safety gain-sharing payouts

state-of-the-art safety procedures and guidelines, Eastern needs to enforce safety and health rules and regulations *consistently* (Table 9.12).

Strategy 5—Build a responsible and knowledgeable workforce. By increasing the skills and abilities of individuals, teams, and supervisors and empowering them, Eastern will be able to increase productivity, reduce operating costs, solve problems, and increase teamwork across the entire plant. Initiatives in support of this priority include training and developmental opportunities in support of self-directed work teams (Table 9.13).

Strategy 6—Improve technology and plant equipment to produce products more efficiently and more economically. Eastern is managing several capital improvement projects to upgrade the condition of equipment and work processes at the plant. Eastern needs to continue these improvements while also employing sound capital project management skills. The company will work to

TABLE 9.13 Strategy 5: Build a Responsible and Knowledgeable Workforce

What we will do	How we will measure achievement
Develop or continue:	Better communication and coordination within team members as measured by regular employee surveys
Empowered, self-directed work team development (decision-making and responsibility)	Innovative, timely, and sound employee and team decision-making as measured by efficiency of meetings and performance results
New performance appraisal system for salaried employees	Better use of tools, equipment, and raw materials, as measured by tool replacement costs
Development planning	Employees are prepared to assume new responsibilities measured by increase in voluntary assignments
Knowledge and skills training for bargaining unit employees	Enhanced partnership agreement measured by reduced grievances
Supervisory development program	Increase in number of ideas and suggestions from employees
Strategic plan communication process	Reduce man-hours per ton
Conversion to parent salary structure	
New bargaining unit job classification (stemming from the labor contract)	
Offer developmental opportunities to sustain employee education and growth through:	Successful development planning
Mentoring	Track progress through training records
Inside training	Enhanced employee performance
Outside technical managerial training	
Opportunities to manage	

TABLE 9.14 Strategy 6: Improve Technology and Plant Equipment to Produce Products More Efficiently and More Economically

What we will do	How we will measure achievement
Complete capital program and budgets each year Conduct major maintenance projects and overhauls	Completion of capital improvements, including: Conversion of production line to point feed technology Substation life extension Cast house continuous homogenizing furnace Rod shop anode cleaner Ladle shop ladle cleaner Bake oven rebuild Production line capacity expansion Rebuilt remelt furnace Developed stack filter systems for metal treatment Facilities expansion Completed stamper upgrades for billet and slab
Conduct research to ensure that Eastern adapts or incorporates improved or emerging technologies	Completion of R&D projects within budget
Develop a stronger capital project management system through training and other developmental assignments	Improved capacity to complete projects on time within budget and schedule

speed up completion of these capital projects and to keep them within budget and quality requirements (Table 9.14).

Strategy 7—Improve Eastern's impact on the environment. Through voluntary efforts and in compliance with state environmental standards, the company will continue to monitor its impact on the local environment. These efforts will be directed toward reducing environmental degradation and pollution (Table 9.15).

TABLE 9.15 Strategy 7: Improve Eastern's Impact on the Environment

What we will do	How we will measure achievement
Comply with federal, state, and local environmental regulations by:	Eliminate incidents of noncompliance
Providing proactive assistance to regulators	Reduce response time for identifying and fixing violations
Educating employees about regulatory requirements	Trainning feedback
Promptly reporting noncompliance and correcting any violations	
Filing a Title V air permit application for environmental capital improvements	Compliance
Participate in voluntary activities on environment, safety and health issues, such as EPA Greenlights, reducing greenhouse gases and PFCs, and noise nuisance reduction	Eliminate environmental, safety, or health complaints about the plant or its operations
Encourage environmentally sound industrial and agricultural growth	Community satisfaction, as measured by community recognition of Eastern's practices
Continue farm production	Farm production and maintenance of safe environmental practices

TABLE 9.16 Strategy 8: Reduce Waste and Non-Value-Added Costs

What we will do	How we will measure achievement
Involve QUEST teams in identifying and resolving quality problems in key production processes	Reduction of rework and scrap tracked by the department on a monthly basis Zero increase in budget for rework and scrap costs
Minimize equipment damage by educating employees, monitoring equipment use, and enforcing rules for properly using equipment	Review monthly maintenance to ensure departmental accountability for responsible equipment use Stay within approved budget guidelines for equipment expenses
Eliminate duplication of effort in administrative processes Process improvement/reengineering Encourage employee use of best practice techniques	Benchmark other processes Monitor process costs
Improve inventory management of supplies and equipment (includes maintenance, production, and raw material in-process)	Reduce inventory by at least 5 percent
Minimize waste generation and increase recycling	Waste product reductions

Strategy 8—Reduce waste and non-value-added costs. Eastern continues to experience waste and non-value-added costs, such as safety and health costs related to accidents, rework and scrap, and equipment damage. Process improvement and problem-solving teams will continue to focus on reducing these costs (Table 9.16).

Postscript to the Strategic Plan

The Eastern strategic plan had been designed as a guidepost for the future, a way of realizing the vision of becoming more responsive to changes going on globally, more supportive to customers and employees, and more cost-effective in manufacturing processes. However, it was not a "cookbook" for success. They recognized that management and employees would continue to have to make informed judgments together each day to make the plan work. And they would have to learn better from their successes and mistakes.

Acquisition and Merger

Although the strategic planning and risk-reduction process was a focused and comprehensive process, and had measurable impacts on plant productivity and success in dealing with its costs and product problems, a major development was not anticipated in the process—acquisition.

During the process of developing and implementing the plan, the company was acquired by a competing parent company, creating a high degree of uncertainty and disruption in the process. Work in implementing the plan was halted until the acquisition and merger process was completed, thus tempering what payoffs could have been produced.

Chapter

10

Risk Lessons Learned and the Project Risk Audit

There are two kinds of "postmortem" on a project and both open up opportunities to look back at the risk management process to see what worked and what did not. This chapter addresses how to do a risk "lessons learned" review and a project risk audit, gives examples, and provides insights on corrective actions based on actual risk feedback from postmortem sessions.

In Fig. 10.1, "lessons learned" is the first step. This is the process of bringing the team together for an informal discussion of what went wrong and what went right in the project and what the impacts were. Then the team identifies where such lessons can be applied in the risk management process, and identifies possible institutional or organizational problems associated with the lessons learned. Potential audit issues are identified and then an audit is conducted on the issues top management would like to earmark for policy, system, or organizational change.

The value of a timely postmortem "lessons learned" review lies in the capturing of insights of the project team members and stakeholders and documented information on the project cycle, fresh from "combat." The process is like a military debriefing in the sense that it is well-focused and tries to capture the intensity of feeling and information all at once. The implications for successful risk management in future projects are major since it is in the insights about what did and did not work that the organization and management team gains valuable information on future project risks and opportunities.

Project Audits

Project audits are not the same as lessons learned. Project audits are performed by external teams who have not been part of the project process—objective outsiders. The audit has been described as "the process of coming down off the

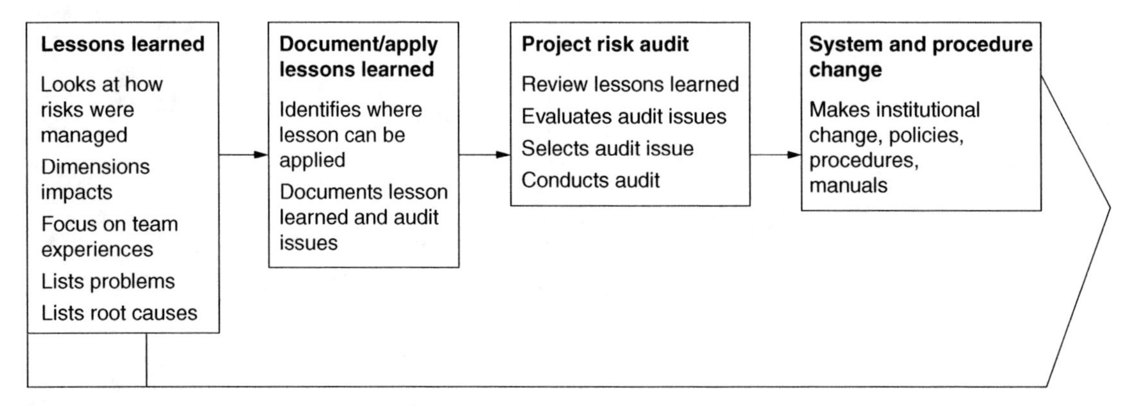

Figure 10.1 Transition from lessons learned to audit.

mountain after the battle to kill the wounded." Such audits can be very useful if they address issues already identified by the project team. The purpose of an audit is to connect factors that contributed to project successes and failures based on documents and recorded information, and to match the project outcomes with project goals and objectives.

How to Do Risk Lessons Learned Review

A lessons learned project review should be conducted soon after project close out and should include all project team members, a member of top management, and stakeholder representatives. The focus should be on identifying what went right and what went wrong and dimensioning what went wrong in terms of unanticipated or unmitigated risks and uncertainty. This can be accomplished by scheduling and facilitating the meeting around key risk issues or topics that help to capture the risks inherent in the project. The following is an example of an outcome report from a risk lessons learned session in a modern avionics instrument product development process.

Example of a lessons learned report

What went right? (Things that we want to recreate for future projects)

- This was a focused team with largely full-time assignments
- Our team was autonomous—we drew the line with the customer appropriate "windows" for changes, disruptions, and the like
- Little or no scope creep
- Resources (e.g., test equip) allocated to resolve problems quickly
- Contingency plans in place when necessary

- Team defined its approach to documentation, and the like together, "as we went" (also noted as a weakness below)
- Communication within team was good
- Consistent high priority on this project from corporate—this was a high-visibility, high-priority project from the beginning and we knew it
- Responsiveness of team members to each other very effective
- Project itself was not technically insurmountable; success was feasible
- Team able to separate what was controllable from what was not—such as flight test—and managed accordingly
- Team was highly proficient; good professional and technical skills
- Scheduling process handled resource issues somehow
- Program manager was open to change; flexible in responding to team issues
- Program manager used schedules as guides but was very task and action oriented in meetings

What went wrong? (Things that we want to correct for future projects)

- Company process and procedure requirements and differing interpretations of document requirements sometimes acted as barriers to necessary work and did not always facilitate successful completion of the project, e.g., software documents, reference documents, and tables
- We sometimes described procedures *after* we completed them instead of before; documentation often followed verification rather than guiding it, e.g., STD checklist
- Confusion and uncertainty in the actual application of ISO, procedures on the one hand, versus "actual" procedures that the team decided were necessary to get the product out on time
- Company changes (split) created resource issues; we had to "ad hoc" it in handling engineering change notices, assembly drawings, and the like because of resource problems created by the split and loss of staff
- Document numbering system created a lot of tension and uncertainty
- Training needs—staff involved in the project were not always trained to carry out procedures, e.g., staff loading the boxes did not have good guidance and training
- Ineffective version control on some documents and configurations
- Stress created by long hours was a problem—can't stretch people and expect them to stay on
- Schedules were not accurate, in many cases, compared to the real work, e.g., the sequence of tests and dry runs

- Sometimes the team did not have the "big picture" on the project; sometimes the big picture helps to facilitate doing your job
- Wasted time in some meetings—meetings had agendas, but there were times when the team "blew off steam" and wasted a lot of time
- Some residual issues rooted in "military" versus "commercial" approach had an impact on the project, which came right in the middle of the transition from military to commercial ways of doing things

Contingency actions

Issue 1. Document numbering system
Recommendation. Set aside time and develop a new document numbering scheme that reflects the way we want to do business

Issue 2. Big gap in documenting actual procedures and processes; problems in sequence and review of documents
Recommendation. Develop and document how we followed or created procedures for ISO audit (they will want to see how we followed our own procedures); review current processes for sequence and review of documents

Issue 3. No accountability for master charting function
Recommendation. Decide clear accountability for master chart function

Issue 4. Don't have the right tools to accomplish process definition, control, and documentation
Recommendation. Decide what the right tools to facilitate documentation are and acquire them, e.g., *Framework* replaces *Word, Agile* is used to control versions, and *Requisite Pro* documents requirements

Issue 5. Ineffective management of document and procedure revisions
Recommendation. Develop a referenced master list of documents and acronyms to ensure effective revision management

Issue 6. Some staff did not follow established procedures and processes
Recommendation. Develop staff accountability for following established procedures and documentation requirements; follow-up with consequences if necessary

Issue 7. Inefficient acquisition of needed equipment
Recommendation. Plug required test equipment early into the scheduling process; anticipate asset issues before they happen

Issue 8. We don't capture observations, problems, and corrective actions along the way, lessons learned are lost
Recommendation. Establish a way to identify and capture project issues, processes, and corrective actions along the way

Issue 9. Conflicts in doing product development in a manufacturing environment; can't build the same thing twice
Recommendation. (this issue was tabled)

Issue 10. There is a widely held view—that we need to turn around—that the company is good at identifying weaknesses but does not follow-up to fix them—there is a concern that issues such as those coming out of this session will not be addressed because of the pressure of time and work.

Recommendation. Make presentation to top management to get personal commitments from key executives on corrective action to mitigate risks on future projects.

A postscript to lessons learned

Risk management is in the end a people-centered process, and it is in the key decisions that people make daily that the conditions for effective risk reduction and response are created. Because the lessons-learned process focuses on the people who actually did the project work and gleans from them a realistic and practical perspective of risks as they played out in the project, the lessons-learned process can be very valuable.

Project Audit

The project audit starts with the appointment of an auditor and an audit team. The auditor is typically another project manager who assumes the role of an auditor with some experience in project planning and control.

The audit, unlike the lessons-learned session, is focused on an independent gathering of information and documents from the project, and reviewing them against the goals and objectives of the project and best practice criteria. This involves the question, "did the project produce what it intended to produce, and how effectively and efficiently?"

Project audit focuses on key aspects of the project, as follows:

1. *Business planning.* Were the risks which actually occurred and which impacted the project identified adequately in the early business planning process?
2. *Follow-up response.* Were those same risks monitored and controlled?
3. *Organization-wide culture.* Did top management support effective risk management as part of the project cycle?
4. *Project team.* Did the project team members perform their roles as "risk managers" during the process and integrate risk into their daily work?
5. *Risk identification, assessment, and response.* Was there a systematic process to identify, assess, and respond to risks?
6. *Key processes, decisions, and milestones.* Were key risk-related processes, such as testing and reliability, quality assurance and control, decision trees, and product development integration milestones, actually followed?
7. *Resources.* Did the project experience resource constraints and were the constraints managed with buffers?

8. *Safety and reliability.* Were the correct tests and reliability processes in place?
9. *Risk-based scheduling.* Were project schedules adjusted using risk inputs and using the MS Project PERT analysis tool?
10. *Monitoring.* Were risks followed in the project process and decisions made on risk mitigation and contingency that reduced risk?

The question of efficiency is reviewed through earned value and cost variance calculations. The issues would be:

Did the project stay consistent with the schedule and budget?

Did the project manager make adequate adjustments based on variations from risk events?

Did the project make its quality, schedule, and budget goals?

Appendix

A

Cost and Risk Exercises—How Do Risk and Cost Work in a Real Project Setting?

The purpose of this appendix is to provide some risk and cost scenarios and questions and answer them so that we can add some reality and practice cases and solutions into this discussion of risk management.

Let's say that you are a project manager in your company and have been assigned to a high-profile, mission-critical project that has high-revenue potential in a new market segment. After an initial assessment, you have determined that it is a risky project since it relies heavily on new technology. However, your upper management does not fully understand and appreciate the basic concepts and approach to project risk management. As a result, you decide to prepare a brief tutorial for your upper management based on the questions that follow.

(a) What is the difference between risk and uncertainty?

SOLUTION: For risk, the probability is known. For uncertainty, the probability is unknown.

(b) According to the PMI Project Management Body of Knowledge, what are the six subprocesses associated with the risk management process?

SOLUTION:

1. Risk management planning
2. Risk identification
3. Qualitative risk analysis
4. Quantitative risk analysis

5. Risk response planning

6. Risk monitoring and control

(c) How do you determine the risk event status?

SOLUTION: Risk event status = risk probability × amount at stake

(d) Once risk has been identified and analyzed, what are the four risk response strategies that may be used to address risk?

SOLUTION: Avoidance, transference, mitigation, and acceptance

Sample Questions

This section contains questions (and answers) that might be used in setting exams for training the class. They are organized by key emphasis areas.

Emphasis area A

Given a project work breakdown structure (WBS), identify the appropriate level for estimating the project, identify the possible types of cost estimates, and recommend the most appropriate alternative.

A1: Types of cost estimates. Since cost is a major area of risk, cost estimates are risks. Analyze the following and recommend the type of cost estimate (*order of magnitude, budget, or definitive*) that should be used to satisfy each of the following circumstances.

(a) Used for project funding and alternative scheme evaluation, when historical information is available.

SOLUTION: Budget

(b) Data include fairly complete plot plans and elevations, equipment data sheets, quotations, and a complete set of specifications.

SOLUTION: Definitive

(c) Needed to support cost control during project execution.

SOLUTION: Definitive

(d) An estimate prepared with the use of flow sheets, layouts, and equipment (but not design) details.

SOLUTION: Budget

(e) An estimate using scale-up or scale-down factors, or an approximate ratio estimate.

SOLUTION: Order of magnitude

A2: Delphi method. As a project manager, you have pulled together a team of subject-matter experts to estimate the task durations for several key tasks on a software development project. You have asked each of them to provide an optimistic, pessimistic, and most likely estimate, assuming that one person would be working on each task (i.e., 40 hours per week, 52 weeks per year). The following are the data that resulted for one of the key tasks:

	Individual estimates (in months)		
Person	Optimistic	Most likely	Pessimistic
Expert 1	1	3	12
Expert 2	2	6	18
Expert 3	3	6	16
Expert 4	3	10	13
Expert 5	1	4	24
Expert 6	2	5	20
Expert 7	5	6	23

(a) Summarize the basic steps in the estimating process using the Delphi method with input from multiple sources (i.e., estimators).

SOLUTION:

1. Assemble estimators in one room
2. Identify the task to be estimated
3. Each person provides an optimistic, pessimistic, and most likely estimate
4. Individual estimates are displayed to everyone in the room
5. Each person discusses his or her assumptions and the issues that he or she considered
6. Individuals adjust their individual estimates based on the discussion
7. Outliers in the final estimates are identified and discarded
8. Averages are calculated for the optimistic, pessimistic, and most likely estimates

(b) Based on the table above, which estimates would you consider outliers that should be discarded?

SOLUTION: Expert 7's optimistic estimate of 5, and Expert 4's most likely estimate of 10 are clear outliers and should be discarded. Although the pessimistic estimates have a wide range, there is no clear outlier and none of these estimates should be discarded.

(c) Calculate the expected duration of the prior task.

SOLUTION:

$$\text{Average (optimistic)} = (1 + 2 + 3 + 3 + 1 + 2)/6 = 2 \text{ months}$$

$$\text{Average (most likely)} = (3 + 6 + 6 + 4 + 5 + 6)/6 = 5 \text{ months}$$

$$\text{Average (pessimistic)} = (12 + 18 + 16 + 13 + 24 + 20 + 23)/7 = 18 \text{ months}$$

$$\text{Expected duration} = 2 + 4\ (5) + 18/6 = 6.7 \text{ months}$$

(d) What other time factors need to be taken into account to adjust the expected duration so that the final estimate is realistic.

SOLUTION:

Holiday, weekends, and vacation time

Time spent mentoring and coaching other team members

Time spent reviewing and/or inspecting other people's work

Interruptions for phone calls and email

Organizational meetings, project meetings, and administration

"Wait time" for information or assistance from others

"Switch time" between multiple tasks and/or projects

A3: Three-point estimate. A construction project requires the use of a special material in several tasks over the next 9 months. Based on recent billings in the geographic area where the construction is being done, the most likely price for the material is $27 /lb. However, there have been price fluctuations over time based on market conditions and material availability. The estimated lowest price is $20 /lb, and the estimated highest price is $35 /lb.

There is also some uncertainty about how much of the material will be required for the project. Site conditions will affect the amount of material actually needed. The most likely estimate is 1,000 lb. However, as little as 800 lb or as much as 1,300 lb might be required.

(a) What is the expected price of the material?

SOLUTION: 20 + 4(27) + 35 = $27.17 /lb

(b) What is the expected amount of material needed for the project?

SOLUTION: 800 + 4(1,000) + 1,300 = 1,016.67 lb

(c) From what sources (i.e., from who) might you receive optimistic, most likely, and pessimistic estimating information?

SOLUTION: Engineers, cost estimators, vendors, and others who are knowledgeable in the areas that you are estimating.

Emphasis area B

Given two or more project alternatives, compare them using equivalent worth and rate of return methods to determine which alternative is superior from a financial perspective.

B1: External rate of return—Risk exposure in financial analysis. Determine the *external rate of return* (ERR) in percent for the following cash flow stream, assuming a reinvestment rate or MARR of 10 percent. Please draw a cash flow diagram to assist in your calculations.

Time	Net cash flow in dollars	Comments
0	(45,000)	Initial investment
EOY 1	20,000	
EOY 2	(5,000)	
EOY 3	30,000	
EOY 4	20,000	
EOY 4	(2,000)	Net salvage value

Note: EOY = end of year.

SOLUTION:

Future accumulation of recovered monies = Future worth of investment

$$(20{,}000)(F/P,\ 10\%,\ 3) - (5{,}000)(F/P,\ 10\%,\ 2) + (30{,}000)(F/P,\ 10\%,\ 1) + 20{,}000 - 2{,}000 = (45{,}000)(F/P,\ i\%,\ 4)$$

$$(20{,}000)(1.3310) - (5{,}000)(1.2100) + (30{,}000)(1.1) + 20{,}000 - 2{,}000 = (45{,}000)(F/P,\ i\%,\ 4)$$

$$26{,}620 - 6{,}050 + 33{,}000 + 18{,}000 = 45{,}000\ (F/P,\ i\%,\ 4)$$

$$71{,}570/45{,}000 = (F/P,\ i\%,\ 4)$$

$$1.5904 = (F/P,\ i\%,\ 4)$$

i	(F/P, i%, 4)	
12%	1.5735	
x	1.5904	.0169/.1755 = .0963
15%	1.7490	

$$(.0963)(.03) + .12 = .0029 + .12 = .1229$$

$$ERR = 12.29\%$$

B2: Internal rate of return. A small computer manufacturing company is thinking about taking on a new, short-term project to develop a specialized performance monitoring system for an Internet service provider. The following are the expected cash flows over the life of the project.

Time	Cash flow in dollars	Comments
0	(250,000)	Initial investment
EOY 1	86,900	Annual net cash flow
EOY 2	111,300	Annual net cash flow
EOY 3	120,000	Annual net cash flow
EOY 3	(25,000)	Net salvage value when project is closed out and sold

Note: EOY = end of year.

(a) Determine the project's *internal rate of return* (IRR).

SOLUTION:

$$PW(10\%) = -250{,}000 + (86{,}900)(P/F, 10\%, 1) + (111{,}300)(P/F, 10\%, 2) + (120{,}000 - 25{,}000)(P/F, 10\%, 3)$$

$$PW(10\%) = -250{,}000 + (86{,}900)(.9091) + (111{,}300)(.8264) + (120{,}000 - 25{,}000)(.7513)$$

$$PW(10\%) = -250{,}000 + 79{,}001 + 91{,}978 + 71{,}374$$

$$PW(10\%) = -7{,}647 \text{ (negative)}$$

$$PW(8\%) = -250{,}000 + 86{,}900\ (P/F, 8\%, 1) + 111{,}300\ (P/F, 8\%, 2) + (120{,}000 - 25{,}000)(P/F, 8\%, 3)$$

$$PW(8\%) = -250{,}000 + (86{,}900)(.9259) + (111{,}300)(.8573) + (120{,}000 - 25{,}000)(.7938)$$

$$PW(8\%) = -250{,}000 + 80{,}461 + 95{,}417 + 75411$$

$$PW(8\%) = 1{,}289 \text{ (positive)}$$

i	PW	
8%	1,289	
x	0	1,289/8,936 = .1442
10%	– 7,647	

$$(.1442)(.02) + .08 = .0029 + .08 = .0829$$

$$IRR = 8.29\%$$

(b) If the MARR is 10 percent, would you recommend that the project be undertaken?

SOLUTION: Recommend that the project not be undertaken, since the IRR of 8.29 percent is below the MARR.

B3: Replacement analysis. A telecommunications company is currently considering replacing an analog switching system in its service network. The old system has operating and maintenance costs of $80,000 per year and it has an expected life of 20 years, at which time it will have no salvage value. In order to meet increasing customer data service needs, the old system will require an immediate upgradation of its digital access and transmission equipment costing $100,000 if it is kept in service.

A new, digital system will cost $400,000 and it will have annual operating and maintenance costs of $50,000 per year. The company has been offered a $50,000 resale price for the old system, and the new system has a salvage value of $100,000 at the end of the 20-year period. Using an MARR of 15 percent and a before-tax analysis, determine whether or not the company should replace the old switching system. Explain your answer.

SOLUTION:

Years	Old system	New system	New system—old system
0	(100,000)	(400,000) + 50,000	(350,000) – (100,000) = (250,000)
1–20	(80,000)	(50,000)	(50,000) – (80,000) = 30,000
20	0	100,000	100,000 – 0 = 100,000

$$\text{PW} = -250{,}000 + (30{,}000)(\text{P/A}, 15\%, 20) + (100{,}000)(\text{P/F}, 15\%, 20)$$

$$\text{PW} = -250{,}000 + (30{,}000)(6.2593) + (100{,}000)(.0611)$$

$$\text{PW} = -250{,}000 + 187{,}779 + 6{,}110$$

$$\text{PW} = -56{,}111$$

Since the present worth is less than zero, do not replace the switching system.

B4: After-tax cash flow. A major petroleum company is considering a new project that is intended to increase the efficiency in a particular part of its oil-drilling operations. The company's after-tax MARR is 20 percent and the company is subject to an effective tax rate of 40 percent. The information listed below represents the estimated cash flows associated with the project.

	Cash flow	
Time	in dollars	Comments
0	(150,000)	Initial investment
EOY 1	45,000	Annual net BTCF
EOY 2	48,000	Annual net BTCF
EOY 3	50,000	Annual net BTCF
EOY 4	55,000	Annual net BTCF
EOY 5	60,000	Annual net BTCF
EOY 6	75,000	Annual net BTCF
EOY 6	20,000	Net salvage value when project is closed out and sold

Note: EOY = end of year, BTCF = before-tax cash flow, ATCF = after-tax cash flow.

(a) Calculate the annual depreciation deductions in the following table:

Year	Factor	Depreciation deduction (in dollars)	Year	Factor deduction	Depreciation deduction (in dollars)
1	.1000	________	7	.0655	________
2	.1800	________	8	.0655	________
3	.1440	________	9	.0656	________
4	.1152	________	10	.0655	________
5	.0922	________	11	.0328	________
6	.0737	________	12	—	—
				Total	150,000

SOLUTION:

Year	Factor	Depreciation deduction (in dollars)	Year	Factor deduction	Depreciation deduction (in dollars)
1	.1	15,000	7	.0655	9,825
2	.18	27,000	8	.0655	9,825
3	.1440	21,600	9	.0656	9,840
4	.1152	17,280	10	.0655	9,825
5	.0922	13,830	11	.0328	4,920
6	.0737	11,055	12	—	—
				Total	150,000

(b) Determine the present worth of the after-tax cash flows associated with the project. Would you recommend that this project be undertaken?

SOLUTION:

Time	BTCF	Depreciation deduction	Taxable income	Income taxes	ATCF	PW(20%)
0	(150,000)				(150,000)	(150,000)
EOY 1	45,000	15,000	30,000	12,000	33,000	27,499
EOY 2	48,000	27,000	21,000	8,400	39,600	27,498
EOY 3	50,000	21,600	28,400	11,360	38,640	22,361
EOY 4	55,000	17,280	37,720	15,088	39,912	19,250
EOY 5	60,000	13,830	46,170	18,468	41,532	16,691
EOY 6	95,000	55,290	39,710	15,884	79,116	26,496

Note: EOY = end of year.
PW(20%) = −10,205.

Since the present worth is negative, do not undertake the project.

Emphasis area C

Given a project cost estimate, recommend a project budget, justify any contingency funds, and recommend a plan for the control of budgeted and contingency spending throughout the project duration.

C1: Project budgeting process. Describe five of the seven identified steps associated with the project budgeting process.

SOLUTION:

1. Get understanding of what client wants
2. Identify work to produce what is wanted
3. See if personnel are available
4. Identify risk involved with doing work
5. Get feedback on amount of time and resources
6. Identify problem that could interrupt work
7. Calculate and publish time and cost target

C2: Project budgeting approaches. Single-project budgeting must fit into the overall budgeting process of the organization and the company as a whole. As a project manager for one of your company's projects, you have been asked to provide a brief description of the top-down and the bottom-up approaches to the budgeting process.

(a) Explain the three basic steps in the top-down budgeting approach by identifying the organizational level and type of budget prepared at each step.

SOLUTION:

Step	Organizational level	Type of budget prepared
1	Upper management	Strategic budget based on organizational goals, constraints, and policies
2	Functional managers	Mid-range budget for each functional unit
3	Project managers	Detailed budgets for each project, including the cost of labor, material, capital equipment, subcontracting, overhead, contingencies, and the like

(b) Explain the four basic steps in the bottom-up budgeting approach by identifying the organizational level and type of budget prepared at each step.

SOLUTION:

Step	Organizational level	Type of budget prepared
1	Upper management	Setting goals and selection of projects (i.e., a framework for budget)
2	Project managers	Detailed budget proposals for projects, including the cost of labor, material, capital equipment, subcontracting, overhead, contingencies, and the like
3	Functional managers	Mid-range budget for each functional unit
4	Upper management	Adjustments and approval of the aggregate budget resulting from the process

Emphasis area D

Given a project description, identify the risks inherent in the project, prioritize these risks, develop strategy for their mitigation or elimination, and communicate this information to management.

D1: Risk management concepts. You are a project manager in your company and have been assigned to a high-profile, mission-critical project that has high revenue potential in a new market segment. After an initial assessment, you have determined that it is a risky project since it relies heavily on new technology. However, your upper management does not fully understand and appreciate the basic concepts and approach to project risk management. As a result, you decide to prepare a brief tutorial for your upper management based on the questions that follow.

(a) What is the difference between risk and uncertainty?

SOLUTION: For risk, the probability is known. For uncertainty, the probability is unknown.

(b) What are the six subprocesses associated with the risk management process?

SOLUTION:

1. Risk management planning
2. Risk identification
3. Qualitative risk analysis
4. Quantitative risk analysis
5. Risk response planning
6. Risk monitoring and control

(c) How do you determine the risk event status?

SOLUTION: Risk event status = risk probability × amount at stake

(d) Once risk has been identified and analyzed, what are the four risk response strategies that may be used to address risk?

SOLUTION: Avoidance, transference, mitigation, acceptance

D2: Risk response strategies. Explain what is meant by each of the following risk response strategies, and give at least two examples for each strategy:

1. Avoidance
2. Transference
3. Mitigation
4. Acceptance

SOLUTION: *Avoidance* is defined as changing the project plan to eliminate the risk or the condition to protect the project goals and objectives from its impact. Examples include:

1. Do not use unfamiliar subcontractors
2. Reduce scope to eliminate high-risk activities
3. Add resources or time to critical tasks during planning
4. Use familiar approaches rather than innovative ones

Transference involves shifting the consequence of a risk to a third party, together with ownership of the risk response. Transferring a risk gives someone else the responsibility for its management. It does not eliminate the risk. Examples include:

1. Use of insurance and performance bonds
2. Use of warranties and guarantees
3. Use of contracts to transfer liability
4. Use of a fixed-price contract with subcontractors

Mitigation reduces the probability and/or consequences of an adverse risk to an acceptable level. Taking early action to reduce a risk's probability and/or impact is more effective than reacting later. Examples include:

1. Adopt less complex processes
2. Plan for additional testing of complex elements
3. Use a more reliable or more stable vendor
4. Use a prototype in the development process

Acceptance means that the project has decided not to change the project plan to deal with a risk or is unable to identify any other suitable response strategy. Examples include:

1. Develop contingency plans
2. Identify risk-trigger points
3. Periodic review of risks and trigger points
4. Use contingency allowance (e.g., time, budget, staff)

D3: Risk identification and response. Your project team has just been assembled to begin preparing a plan for the following project:

1.0 Conducting Seminar

- 1.1 Project Management
  - 1.1.1 Planning
    - 1.1.1.1 Scheduling
    - 1.1.1.2 Risk Management
  - 1.1.2 Administration
  - 1.1.3 Budget
- 1.2 Marketing Preparation
  - 1.2.1 Brochures
    - 1.2.1.1 Planning
    - 1.2.1.2 Printing
    - 1.2.1.3 Labeling
    - 1.2.1.4 Mailing
- 1.3 Material Preparation
  - 1.3.1 Design
  - 1.3.2 Printing
    - 1.3.2.1 Reproduction
- 1.4 Conduct Seminar
  - 1.4.1 Reservation
  - 1.4.2 Registration
  - 1.4.3 Presentations
  - 1.4.4 Critique Sheets
- 1.5 Logistics and Support
  - 1.5.1 Transportation
  - 1.5.2 Accommodations
  - 1.5.3 Meals

Your project sponsor wishes to get an initial reading (based on the 3 to 4 level WBS that has been provided) on what risks are involved and how they should be managed. Explain at least five risk areas associated with the project and provide a brief response plan for each.

SOLUTION:

Project risk area	Risk response plan
1. Presenter availability	1. Use alternate presenters
2. Presenter availability	2. Plan back-up presenters
3. Printer errors	3. Product checkpoints in plan
4. Printer errors	4. Contractor performance requirements
5. Insufficient accommodations for clients	5. Contract for back-up rooms
6. Late registrations	6. Space contingencies
7. Cost overruns in materials	7. Contingency fund
8. Presentation material errors or lost	8. Contract with local quick minute changes copy service

D4: Evaluation of risk events or opportunities (16 points). You are the project manager for a firm constructing a new electronic manufacturing facility. The facility is designed to last 15 years. The initial capital cost to construct the plant is estimated to be $150,000,000. In conducting an environmental scan of factors affecting your project, you have identified a 10 percent chance that litigation by environmental groups over land use could delay the start of construction by 2 years and add $25,000,000 to the cost of construction.

Over the useful life of the facility, the plant is expected to manufacture components that are expected to yield $40,000,000 each year in net profits. Further, your marketing department has forecast that there is a 25 percent chance that heightened demand for the components may allow you to raise prices and realize profits as much as $50,000,000 per year of operation. However, there is also a possibility of a new competitor emerging in the near future. While you view this as unlikely, with only a 10 percent probability, if it does occur it will decrease your firm's profits by $10,000,000 per year due to the heavy price competition that would ensue.

However, you are aware from your environmental scan that the local government may raise real estate taxes to pay for road improvements and the like because of the explosive growth the area has experienced recently. Based on the analysis by your firm's tax lawyers, you estimate that if the tax increase goes through, it will decrease your net profits by $2,000,000 per year. You estimate there is a 40 percent chance of this occurring.

The manufacturing of electronic components uses some very volatile chemicals. Although your firm has extensive experience in manufacturing with these facilities, you know from historical data that there is a 20 percent chance of a chemical spill in any year of operation. The occurrence of such a spill would decrease your firm's operating profits by $2,000,000 per year. Insurance is available to protect you from such losses at a cost of $250,000 per year.

As part of the proposal, and in order to achieve all the necessary permits, your firm has agreed to clean the facility at the end of its useful life and convert the facility for an unspecified, nontoxic, government use. You estimate that this will cost $40,000,000. However, you also estimate there is a 30 percent chance that these costs could increase by 25 percent by the end of the plant's life due to increased environmental regulation.

Assume that all costs are in normalized dollars, i.e., *do not* discount future cash flows.

(a) What are the opportunities and risk events presented in the scenario?

SOLUTION: Litigation, price increase, competition, tax increase, chemical spill, and regulatory impacts

(b) What is the value of the project if none of the risk events or opportunities occur?

SOLUTION:

$$\text{Value} = -\$150{,}000{,}000 + (15 \text{ years})(\$40{,}000{,}000 \text{ per year}) - \$40{,}000{,}000$$

$$\text{Value} = \$410{,}000{,}000$$

(c) What is the value of the project, including all risk events and opportunities?

SOLUTION:

$$\text{Value} = \$410{,}000{,}000 - (0.1)(\$25{,}000{,}000) + (0.25)(\$150{,}000{,}000) - (0.1)(\$150{,}000{,}000) - (0.4)(\$30{,}000{,}000) - (0.2)(\$30{,}000{,}000) - (0.3)(\$10{,}000{,}000)$$

$$\text{Value} = \$410{,}000{,}000 - \$2{,}500{,}000 + \$37{,}500{,}000 - \$15{,}000{,}000 - \$12{,}000{,}000 - \$6{,}000{,}000 - \$3{,}000{,}000$$

$$\text{Value} = \$409{,}000{,}000$$

(d) What is the value of the project if all risk events or opportunities occur in their worst case?

SOLUTION:

$$\text{Value} = \$410{,}000{,}000 - \$25{,}000{,}000 - \$150{,}000{,}000 - \$30{,}000{,}000 - \$30{,}000{,}000 - \$10{,}000{,}000$$

$$\text{Value} = \$165{,}000{,}000$$

(e) What is the value of the project if all risk events or opportunities occur in their best case?

SOLUTION:

$$\text{Value} = \$410{,}000{,}000 + \$150{,}000{,}000$$

$$\text{Value} = \$560{,}000{,}000$$

Emphasis area E

Given a project description and identified risks, perform a sensitivity analysis to determine the effects of identified risks on the project.

E1: Sensitivity analysis. It is necessary to determine the number of lanes on a short tollway extension project. It is expected that the tollway will be in service for 30 years and then be demolished with a net zero salvage value. Following are pertinent data for the initial cost, annual *operating and maintenance* (O& M) expenses, and annual revenues from tolls based on two scenarios:

Description	Four lanes	Five lanes
Initial cost	$56,000,000	$72,000,000
Annual O&M expenses	$2,300,000	$2,960,000
Annual revenues (optimistic)	$10,000,000	$12,500,000
Annual revenues (pessimistic)	$8,900,000	$11,525,000

Analyze the sensitivity of the decision due to the optimistic and pessimistic estimates of annual revenues by using the *present worth* (PW) method. Assume an MARR of 10 percent. Recommend the number of lanes that should be constructed. Explain your answer.

SOLUTION:

$$\text{PW(4 opt)} = -56{,}000{,}000 + (10{,}000{,}000 - 2{,}300{,}000)(P/A, 10\%, 30)$$

$$\text{PW(4 opt)} = -56{,}000{,}000 + (7{,}700{,}000)(9.4269) = \$16{,}587{,}130$$

$$\text{PW(4 pess)} = \$6{,}217{,}540$$

$$\text{PW(4 range)} = \$10{,}369{,}590$$

$$\text{PW(5 opt)} = -72{,}000{,}000 + (12{,}500{,}000 - 2{,}960{,}000)(P/A, 10\%, 30)$$

$$\text{PW(5 opt)} = -72{,}000{,}000 + (9{,}540{,}000)(9.4269) = \$17{,}932{,}626$$

$$\text{PW(5 pess)} = \$8{,}741{,}398$$

$$\text{PW(5 range)} = \$9{,}191{,}228$$

Since PW(5) is higher in both the optimistic and pessimistic cases and the range (optimistic–pessimistic) of PW(5) is lower, select the 5-lane alternative.

E2: Break-even point and payback period. A project is currently under consideration by a company, but uncertainty exists regarding the life of the project. The following are the most likely (or expected) estimates for the project:

Initial investment	$125,000
Project life	5 years
Salvage value	$5,000
Annual receipts	$90,000
Annual disbursements	$47,000
MARR	8%

(a) Using the *annual worth* (AW) method, determine the break-even point of the project in years.

SOLUTION:

$$\text{Annual net cash flow} = (P - S)(A/P, i\%, n) + (S)(i\%)$$

$$(90{,}000 - 47{,}000) = (125{,}000 - 5{,}000)(A/P, 8\%, n) + (5{,}000)(.08)$$

$$43{,}000 = (120{,}000)(A/P, 8\%, n) + 400$$

$$(A/P, 8\%, n) = .3550$$

n	A/P	
3	.3880	
x	.3550	.0330/.0861 = .38
4	.3019	

Break-even point = 3.38 years

(b) Based on the result in part (a), would you recommend the project for acceptance?

SOLUTION: Since break-even is 3.38 years and project life is 5 years, recommend a "go ahead."

(c) Calculate the payback period (in years) for this project and explain why it is different from the break-even?

SOLUTION: Simple payback = 125,000/43,000 = 2.907 years

The break-even recognizes the time value of money, the payback does not.

E3: Break-even and sensitivity. A major communications carrier is faced with increasing demand on its long-haul transmission facilities between Chicago and St. Louis. As a result, the company is considering a new project to expand its transmission facilities on that route. There are two project alternatives: (1) install equipment with full transmission capacity now, or (2) install equipment in two phases. The following table shows the estimated installation costs:

Alternative	Cost
Full-capacity Installation	$140,000
Two-phase installation	
Install first phase now	$100,000
Install second phase n years from now	$120,000

All facilities will last until 40 years from now regardless of when they are installed. At that time, they will have no salvage value. The annual cost of operating and maintaining the facilities is the same for both alternatives. The company's MARR is 8 percent at this time.

(a) Using the present worth (PW) method, plot a simple graph showing "year second phase installed" (i.e., the x-axis) versus "PW of cost" (i.e., the y-axis) for both alternatives. From the graph, determine the break-even point in years.

SOLUTION: The x-axis should range from 0 to 30 years. The y-axis should range from 0 to $250,000.

Full-capacity installation alternative:

PW of cost = $140,000, so this alternative is represented by a horizontal line.

Two-phase installation alternative:

PW of cost = $100,000 + ($120,000)(P/F, 8%, n)

Calculate PW of cost for n = 0, 5, 10, 20, and 30 years

n = 0	PW of cost = 100,000 + 120,000 = $220,000
n = 5	PW of cost = 100,000 + (120,000)(0.6806) = $181,700
n = 10	PW of cost = 100,000 + (120,000)(0.4632) = $155,600
n = 20	PW of cost = 100,000 + (120,000)(0.2145) = $125,700
n = 30	PW of cost = 100,000 + (120,000)(0.0994) = $111,900

This alternative is represented by a plot of five points that decrease from left to right.

Break-even point = 15 years (approximately, from the graph)

(b) If it is estimated that the second-phase capacity will be needed in year 18, how sensitive is the decision to this estimate?

SOLUTION: If the second-phase capacity would be needed over the next 5 to 10 years, the decision is not sensitive to this estimate.

If the second-phase capacity would be needed between years 12 and 18, the decision is sensitive and would depend on the estimate of when full capacity would actually be needed.

If the second-phase capacity would be needed in year 18 and beyond, the decision on which alternative to use is not sensitive to this estimate.

Emphasis area F

Given identified project risks, apply expected value and decision tree techniques to evaluate likely project outcomes.

F1: Payoff matrix/expected value. A company is planning a project to develop a new product and has identified three strategies it can use to bring the product to market—A, B, and C. Three possible states of nature are also identified—a low market demand, an even market demand, and a high market demand. A payoff matrix, including the probability of each market condition, has been developed:

	States of nature (profits in $million)		
Strategy	High = 20%	Even = 50%	Low = 30%
A	$50	$20	($30)
B	$70	$40	($60)
C	$25	$15	$10

Determine the expected value of each strategy and recommend what strategy the project should take.

SOLUTION:

Expected value = (payoff high)(probability high) + (payoff even)(probability even) + (payoff low)(probability low)

Expected value of A = ($50)(0.2) + ($20)(0.5) + (–$30)(0.3)

Expected value of A = $10 + $10 – $9 = $11

Expected value of B = ($70)(0.2) + ($40)(0.5) + (–$60)(0.3)

Expected value of B = $14 + $20 – $18 = $16

Expected value of C = ($25)(0.2) + ($15)(0.5) + ($10)(0.3)

Expected value of C = $5 + $7.5 + $3 = $11

For this project, the strategy of choice would be B since it has the highest expected value.

F2: Expected monetary value. One of your projects is estimated to have a variable life between 3 and 7 years. The following project estimates were arrived at using historical data as well as expert judgment:

Initial investment	$10,000
Project life	3 years with probability = 0.3
	5 years with probability = 0.4
	7 years with probability = 0.3
Salvage value	$2,000
Annual receipts	$5,000
Annual disbursements	$2,200

Determine the expected annual worth (AW) of the project if the MARR is 8 percent.

SOLUTION:

For life = 3 years, the net AW is:

$$\$5{,}000 - \$2{,}200 - [(\$10{,}000 - \$2{,}000)(A/P, 8\%, 3) + (\$2{,}000)(.08)] = -\$460$$

For life = 5 years, the net AW is:

$$\$5{,}000 - \$2{,}200 - [(\$10{,}000 - \$2{,}000)(A/P, 8\%, 5) + (\$2{,}000)(.08)] = \$630$$

For life = 7 years, the net AW is:

$$\$5{,}000 - \$2{,}200 - [(\$10{,}000 - \$2{,}000)(A/P, 8\%, 7) + (\$2{,}000)(.08)] = \$1{,}110$$

$$\text{Expected AW} = (-\$460)(0.3) + (\$360)(0.4) + (\$1{,}100)(0.3)$$

$$\text{Expected AW} = \$446$$

F3: Decision tree analysis. Silver will be used in a manufacturing process on a new product development project. You must decide "to stock" or "not to stock" a large supply of silver. The uncertain variable is the future price of the metal. You have three *options*:

1. Stock silver in a large supply

   Future price, probability, and present value of silver stock are as follows:

Price	Probability	Present value
High	.3	$25,000
Medium	.4	$5,000
Low	.3	$–15,000

2. Stock no silver with present value = $0
3. Hire a consultant (at a cost of $3,000) to predict if silver prices will go higher or lower. Then, based on that prediction (estimated to be 60 percent higher and 40 percent lower), determine "to stock" or "not to stock" a large supply of silver:
   - Stock silver in large supply

   Future price, probability, and present value of silver stock are as follows if the prediction is for higher prices:

Price	Probability	Present value
High	.4	$25,000
Medium	.45	$5,000
Low	.15	$–15,000

Future price, probability, and present value of silver stock are as follows if the prediction is for lower prices:

Price	Probability	Present value
High	.2	$25,000
Medium	.3	$5,000
Low	.5	$–15,000

- Stock no silver with present value = $0

(a) Diagram the problem in the form of a decision tree.

(b) Determine what would be the better *option* using the *expected monetary value* (EMV) criterion.

SOLUTION:

Decision 3 (25,000)(.2) + (5,000)(.3) – (15,000)(.5) = –1,000 versus 0

Decision 2 (25,000)(.4) + (5,000)(.45) – (15,000)(.15) = 10,000 versus 0

Decision 1

Stock in large supply = (25,000)(.3) + (5,000)(.4) – (15,000)(.3) = 5,000

No stock = 0

Hire consultant = (10,000)(.6) + (0)(.4) – 3,000 = 3,000

Because 5,000 is greater than 3,000, don't hire a consultant but stock silver in large supply.

Emphasis area G

Given a project in progress, collect data on cost and schedule parameters to determine whether the project is on plan. Use earned value calculations to determine whether control actions are to be taken. Develop recommendations to achieve a satisfactory outcome and describe a reporting plan.

G1: Project status and reporting. As the project manager for a key project your firm is delivering to a top client and you are preparing a progress report to your client indicating your project's performance. The project has been in actual progress for 9 months. The information below provides the original schedule of task durations, budgeted costs, performance to date, and a network diagram.

Task	Duration (months)	Budgeted costs	Percent complete	Actual dollars expended
A	2 months	\$100,000	100%	\$150,000
B	1 month	\$100,000	100%	\$125,000
C	4 months	\$600,000	75%	\$500,000
D	2 months	\$200,000	0%	\$0
E	2 months	\$150,000	0%	\$0
F	1 month	\$50,000	0%	\$0
		\$1,200,000		

Network diagram: A → B → C → D → E → F

(a) Based on this information, prepare a cost/schedule status report to show what you would share with your client.

SOLUTION:

Budget at completion (BAC) = sum of budgeted costs = \$1,200,000

BCWS for the 9-month point = Budget for A + B + C + D = \$1,000,000

BCWP = \$100,000 + \$100,000 + (0.75)(\$600,000) = \$650,000

ACWP = \$150,000 + \$125,000 + \$500,000 = \$775,000

Cost variance = BCWP – ACWP = \$650,000 – \$775,000 = –\$125,000

Cost performance index = BCWP/ACWP = \$650,000/\$775,000 = 0.84

The project is overbudget.

Schedule variance = BCWP – BCWS = \$650,000 – \$1,000,000 = –\$350,000

Schedule performance index = BCWP/BCWS = \$650,000/\$1,000,000 = 0.65

The project is behind schedule.

Variance ACWP/BCWP = \$775,000/\$650,000 = 1.19

Completed tasks = (1.0)(\$150,000) + (1.0)(\$125,000) = \$275,000

In-progress tasks = \$500,000/.75 = \$667,000

Tasks not started = (\$775,000/\$650,000)(\$400,000) = \$477,000

Total *estimate at completion* (EAC) = \$275,000 + \$667,000 + \$477,000 = \$1,419,000

(b) What would be the best way to deliver this report to your client? What other information would you convey at the same time?

SOLUTION: The project is clearly in jeopardy. It is critical that you prepare a formal presentation to your client that includes at least the following.

You should objectively present the current project status. This would be best presented with a graphical depiction of the project status that will be easily understood by the client. You should also present the factors, events, or conditions that brought about this status and how you had anticipated them in your project risk management plan.

You should present the measures and actions that you are implementing to correct or eliminate any further slippage in project performance. The actions that you have already taken, or intend to take in the near future, are the key items to present to your client, considering the current overall project status. This should lead to a current view of the project's estimated completion dates and costs.

G2: Analysis of project data. You have been provided the following information about the status of Task A on your project:

Status of Task A

Item	Budget	BCWS	BCWP	ACWP	CV	SV
Direct labor (hours)	5,000	3,500	3,250	3,000		
Labor cost (dollars)	62,500	43,750	42,500	45,000	(2,500)	(1,250)
Material cost (dollars)	75,000	55,000	54,000	57,000	(3,000)	(1,000)
Labor overhead (dollars)	25,000	17,500	15,500	16,200	(700)	(2,000)
Material handling (dollars)	15,000	11,000	10,000	13,110	(3,110)	(1,000)
Total (dollars)	177,500	127,250	122,000	131,310	(9,310)	(5,250)

(a) Is Task A on schedule? What is the scheduled and the actual completion percent?

SOLUTION:

Based on labor hours: Scheduled = BCWS/Budget = 3,500/5,000 = 70% (Behind)

Based on labor hours: Actual = ACWP/Budget = 3,000/5,000 = 60% (Behind)

(b) What assumptions about costs were made during budget development?

SOLUTION:

Labor rate = ? = \$62,500/5,000 h = \$12.50 /h for budget

Labor overhead = ? = \$25,000/\$62,500 = 40% for budget

Material handling = ? = \$15,000/\$75,000 = 20% for budget

(c) Is the project meeting the targets in part (b) above?

SOLUTION:

Actual labor rate = ? = \$45,000/3,000 h = \$15.00 /h (greater than budget)

Actual labor overhead = ? = \$16,200/\$45,000 = 36% (less than budget)

Actual material handling = ? = \$13,110/\$57,000 = 23% (greater than budget)

G3: Project control devices. Explain three project control devices that could be used in the schedule area, three in the cost area, and three in the quality area.

SOLUTION:

	Area	*Control devices*
1.	Schedule	milestones
	Schedule	*critical path method* (CPM)
	Schedule	summary reports
	Schedule	Gantt charts
	Schedule	earned value measurement
2.	Cost	earned value measurement
	Cost	productivity
	Cost	bid estimates/budgets
	Cost	Contract provisions
	Cost	Change pricing
3.	Quality	design documents
	Quality	code and local requirements
	Quality	testing and inspection
	Quality	industry standards
	Quality	punch lists

Emphasis area H

Given multiple projects that compete for resources within the firm, use capital rationing techniques to determine which projects should be funded.

H1: Capital rationing. The following information has been developed for five independent project investment opportunities that have been identified by your marketing, manufacturing, regulatory compliance, and engineering departments:

		Cash flow (in $thousands) at end of year				
No.	Project area	Year 0	Year 1	Year 2	Year 3	Year 4
1	Marketing	−10	4	6	5	5
2	Manufacturing	−15	8	7	6	2
3	Reg. compliance	−5	1	3	4	5
4	Engineering (A)	−7	4	4	4	4
5	Engineering (B)	−8	3	4	5	6

Calculate the *profitability index* (PI), the payback period, and the net present value (NPV) at 20 percent (the MARR) for each of the above projects. Then, based on *your* selection criteria (such as profitability index, payback period, net present value), recommend how $25,000 of internally generated funds should be spent. Note: Funds may be assigned to future projects.

SOLUTION: Since all projects have a positive net present value, all could be chosen. Decision criteria on each project are:

Project number	Investment (in $thousands)	Profitability index	Payback period (yrs.)	NPV/investment
1	10	1.28	2	.28
2	15	1.06	2	.06
3	5	1.53	2.25	.53
4	7	1.48	1.75	.48
5	8	1.38	2.20	.38

Based on the financial criteria described above, the following recommendations would be made:

Based on PI: approve 3, 4, and 5 with $5,000 left for new projects.

Based on payback: approve 4 and 1 or 2 with funds available *or* approve 1 and 2.

Based on NPV: approve 3, 4, and 5 with $5,000 left for new projects.

H2: Let's ration some capital—what a capital idea! Your budget for projects is $400K. Data for simple Projects A, B and a more complex C are shown below. All have 8-year economic lives, and there are no salvage values.

Project	Initial cost	Annual net revenue
A	$100K	$40K
B	$170K	$50K
C1	$200K	$50K
C2	$310K	$60K

Project A can be undertaken in any combination with the others. However, Projects B and C are mutually exclusive. Further, Project C can actually be considered as two projects where Project C1 would be a small version and Project C2 would be a large version. You could choose C1 or C2, and neither could be paired with B.

(a) Using an NPV framework, ration the $400K budget in the most efficient way. Assume that the MARR is 12 percent.

SOLUTION: We can eliminate C2 right away because if we consider "expansion" of C1 into C2, we would spend $110K up front and generate only $80K in net revenues over the life of the project. Thus, we are left with five options to evaluate—A alone, B alone, C1 alone, A + B, and A + C1.

Calculating relevant NPVs (leaving out the dollar signs), we get:

$$NPV(A) = -100K + 40K(P/A, 12\%, 8) = -100K + 40K(4.97) = \$98.8K$$

$$NPV(B) = -170K + 50K(P/A, 12\%, 8) = -170K + 50K(4.97) = \$78.5K$$

$$NPV(A) = -200K + 50K(P/A, 12\%, 8) = -200K + 50K(4.97) = \$48.5K$$

From the above results, it is obvious that the combination A + B is superior to the others. Why? Because the NPVs for both projects are positive and sum of the NPVs (i.e., $177.3K) is the highest possible.

(b) Speculate in a reasoned way on your choices if you could spend an unlimited amount of money on projects.

SOLUTION: If we had an unlimited amount of funding, I would still choose the combination A + B. The mutually exclusive nature of Projects B and C precludes us from undertaking both and we have shown already that A + B is preferred to A + C.

More Questions on the Risk Management Process and Cost

Work breakdown structure

Questions and examples that focus on the description and purpose of the work breakdown structure and its use in decomposing a project into a set of integrated tasks and activities

Problems that demonstrate WBS levels, work packages, and deliverables

Problems and scenarios that develop the coding of WBS elements and the linkage to cost accounting systems

Estimating

Problems that demonstrate the difference between order of magnitude, budget, and definitive project estimates and their application

Questions that describe the methods of estimating activity durations and estimating various types of costs (e.g., direct, indirect, capital, and the like) and the bases used for project estimating

Scenarios and cases that identify and address major issues associated with estimating

Project financial perspectives

Questions that require the student to perform calculations using interest rate, discount rate, and minimum acceptable rate of return

Problems, scenarios, and cases that require the understanding and use of equivalent worth methods, rate of return methods, break-even analysis, and payback period in the evaluation and selection of project alternatives.

Questions and problems that demonstrate the effects of depreciation and taxes on project alternatives

Project budgeting

Problems and questions that describe the inputs to the project budgeting process, general approaches to budget preparation, and risk considerations

Questions that demonstrate the purpose and application of contingency funds and management reserves to account for risk

Scenarios or cases that describe project financing alternatives and capital rationing techniques to deal with the issue of limited resources

Risk analysis and decision criteria

Problems that require the student to understand risk and risk events, types of project risks, the steps associated with the risk management process, and risk probability and impact

Scenarios and cases that require the student to apply risk impact assessment techniques including expected value, decision trees, sensitivity analysis, expert judgment, and simulation

Questions that apply risk response strategies, as well as monitoring and control mechanisms

Project control systems

Questions that describe the development of the requirements for project monitoring and control—what to monitor, how often, and from what source the data will come; and timeliness of progress/performance reporting.

Problems, scenarios, and cases that address the project issues of scope management, cost/schedule/performance tradeoff, and change management

Cost/schedule management

Problems that require the understanding and computation of earned value measures—earned value at project review points, cost variance, cost performance index, schedule variance, schedule performance index, and project estimate at completion

Scenarios or cases that apply earned value measures to project decision making, recovery alternatives, and replanning efforts

Cost/schedule control systems criteria (C/SCSC)

Questions that focus on the description and background of the cost/schedule control systems criteria (C/SCSC) established for project control, as well as the major areas of the criteria—organization, planning and budgeting, accounting, analysis, and revisions and access to data

Appendix

B

Risk-Based Project Schedule

This appendix is a Microsoft Project schedule showing the results of the PERT analysis. Note the calculated durations based on expected, pessimistic, and optimistic estimates taken from a risk matrix.

PERT Analysis Schedule

ID	Acquire terminals in strategic areas	Duration	Optimistic dur.	Expected dur.	Pessimistic dur.
1	Study population and expansion areas	122.5 days	110 days	155 days	115 days
2	Determine population growth	17 wks	12 wks	16 wks	21 wks
3	Lineup finances	15.33 wks	10 wks	14 wks	20 wks
4	Determine market for terminals	9.17 wks	10 wks	15 wks	0 days
5	Assess terminal worth	14.67 wks	12 wks	14 wks	17 wks
6	Personnel availability	15.33 wks	10 wks	16 wks	17 wks
7	Assess terminals	104.17 days	75 days	100 days	125 days
8	Proper personnel utilized	7.33 wks	8 wks	12 wks	0 days
9	Engineering studies performed	20.83 wks	15 wks	20 wks	25 wks
10	Evaluate equipment	9 wks	6 wks	8 wks	12 wks
11	Environmental hazards limited	8.33 wks	6 wks	8 wks	10 wks
12	Drawings of terminals available	2.17 wks	1 wk	2 wks	3 wks
13	Employees	159.17 days	95 days	110 days	240 days
14	Use new or existing employees	4 wks	2 wks	4 wks	5 wks
15	Accountant for terminal purchase	15.67 wks	10 wks	16 wks	18 wks
16	Legal documents in order	13 wks	10 wks	12 wks	16 wks
17	Financing complete	11.33 wks	8 wks	10 wks	15 wks
18	Documentation finalized	11.33 wks	8 wks	10 wks	15 wks

Appendix

C

Demystifying Business and Project Risk Management: A Checklist

Demystifying Business and Project Risk Management: A Checklist

Action	What?	Why?	When?	Output?	Who?
		Business culture			
Create Risk Management policy	Create business intent to manage risk	Confirm that it is important and back it up	Part of business plan; underlies project process	Policy statement on how the business will handle risk	Executive and program management level
Assess organization awareness	Find out how aware workforce is of risk and risk response impacts	Survey workforce	Every 6 months	Workforce awareness of risk management report	HR/Project team
Deliver training program	Design training around practice planning tools; use to introduce business risk	Workforce will implement if they understand tools	Every year with refresher	Certification	All pm, teams, and technical personnel
Reward effective risk management	Provide rewards for good risk management effort and effectiveness	Incentives motivate	During project	Compensation reward	Project managers and team members
		Business strategy			
Risk component of business plan	Provide for a risk section in the business plan and communicate it	SWOT analysis; threats = risks Translate to product line risk exposure	Annual update of business strategic and business plan	Risk-based business plan; integrate with financial and profitability analysis	Executive and program managements
Strategic objectives	State objectives in terms of risk	Measurable strategy goals	Part of plan; communicate to workforce	Set of 10 long term objectives	Executives and program managers
		Project selection			
Do risk assessment of candidate projects	In developing business portfolio of projects, use risk as one criterion for project selection	Use *PMBOK* process; broad-brush risk assessment	Each time project portfolio pipeline is updated	Rank order projects using composite risk, alignment, cost and revenue assessment	Program managers and functional managers

Weigh risk against revenues and alignment	Demonstrate that risk has been embedded in business and financial analysis	Trade off risk with opportunity for profitability and taking advantage of business core competence	Each time pipeline is updated	Analysis, data, documentation	Project management office, project team, business planning staff
			Project plan		
Requirements	State customer requirements in terms of customer risks	Risk that customer requirements do not reflect risk, or misunderstanding customer perspective and expectations on project risks	During initial concept phase part of project plan	Requirements document stating customer requirements and risks	Project manager functional manger, and customer
WBS	Include risk contingencies from risk matrix in WBS work activity	Because there is inherent risk in missing major parts of the deliverable in initial planning; WBS assures coverage of major "chunks" of work	During development of the deliverable, the "work" should include initial contingencies identified in risk assessment	WBS in organization chart form and outline in MS Project	Project manager
Task list	Include risk tasks and contingencies in baseline schedule	Task list should include all anticipated contingency actions should risk events occur	After WBS is prepared, do task list and link; there is inherent risk that linkages will be too "hard;" allow for "soft" linkage	Task list in MS Project Gantt chart or spreadsheet	Project management office and/or project manager
Network diagram	Show risk in network diagram with 3 scenarios, expected, pessimistic, and optimistic	Arrow diagram shows critical and non-critical paths; risks inherent in focusing on critical path when resource constraints in non-critical tasks may serve as bottleneck theory of constraints	During translation of WBS to Gantt chart, prepared to show dependencies and paths	Arrow diagram in MS Project or other software	Project management office template, or project manager

(*Continued*)

Demystifying Business and Project Risk Management: A Checklist (*Continued*)

Action	What?	Why?	When?	Output?	Who?
		Project plan *(Continued)*			
Calendar based diagram	Relate network to time to begin to see schedule impacts and milestones	Histogram using arrows and calendar	During translation of WBS to Gantt chart	Graphics software or Word document	Project manager
Risk-based schedule	Do risk-based schedule using MS Project PERT analysis tool	MS project Gantt chart showing calculated risk-based duration after weights and 3 scenarios are entered	During initial scheduling, then any time risk is identified and contingency prepared	MS Project schedule file showing calculated durations for high risk tasks	Project management office or project manager
		Risk management process (see *PMBOK*)			
Risk identification	Using input from business plan, identify and rank project tasks in terms of risk	Using data and information and past experience, rank summary tasks in WBS using risk matrix	During business planning, project and portfolio selection, and project WBS scheduling	Risk matrix	Project management office or project manager
Risk assessment	Assess risks using risk matrix format	Complete risk definition, impact (schedule, cost, quality, business growth); make probability estimate (25%, 50%, 75% probability), severity on project outcome, and contingency	During business planning, project selection, and project planning and control	Risk matrix, updated monthly	Project manager
Risk response	Prepare contingency actions and include in baseline schedule	Response is planning through definitive contingency plans and tasks which are embedded in project schedule as regular tasks-triggered if risk event occurs	During project planning, responses and contingencies are designed to address specific risks and recorded; this is where the team anticipates what might happen	Risk contingency actions	Project manager

			to slow or delay the project, what can be done to prevent it or address it, and schedules contingency tasks into the project		
Risk matrix	Prepare risk matrix as basis for scheduling	The risk matrix is the basic checklist item for risk throughout the process; it is the guide for action	During project planning a basic risk matrix file is established and appears with all project planning and project review documents	Risk matrix following prescribed format	Project manager and team or task managers
Decision tree	Do decision tree analysis to expected value of optional decisions	This is the way project managers anticipate decisions they will have to make based on risk, and what alternative paths and expected values will follow each decision path	During project planning, decision tree analysis is applied to high risk tasks	Decision tree diagram with expected values calculated	Project manager
Integrate risk into project manual					
Basic project manual	Assure that risk is not treated separately, but seen as part of the way projects are planned and controlled	Because risk should not be treated separately from project management process; manual captures how risk is integrated into process	Business establishes a system of basic project manuals as part of "projectizing" the organization	Online and hardcopy manual including basic project planning and risk management tools and templates	Project management office (PMO)
Provide software tools	Train and provide software analysis tools in manual	Much of the risk analysis can be done through spreadsheets and decision tree analysis software; workforce needs to know how to use them	Business establishes a support system of risk management application software and trains appropriate staff	Software library	IT and project managers

(Continued)

Demystifying Business and Project Risk Management: A Checklist (*Continued*)

Action	What?	Why?	When?	Output?	Who?
		Product/technical process development			
Define product/technical development process in generic WBS	Assure that business has defined the core product development and technical processes which it uses to produce products and services, e.g., engineering, construction, system development, standardizing where possible	Because project risk management cannot be successful unless both technical and product development result a requirements are conducted to control risk, and management impacts, e.g., schedule and cost are applied to the real industry processes that create customer value	Business establishes a generic WBS of technical processes; these are recorded and updates so that all project WBS and schedule information, and risk data, is taken from the generic model and tailored	WBS file	Functional managers
Project codes	Provide for coding actions in WBS so that costs can be captured	Because once you have identified all tasks and risk contingencies, you will want to capture costs against those codes to build a history of risk management and mitigation costs	When generic WBS is set up, codes are added at the appropriate level to capture costs	Coding system integrated with time sheets and accounting system	Accounting, project management

Identify customer risk tolerance					
Assess customer perspective on business and project risk and how much risk customer is willing to assume	Solicit customer input on customer risks and uncertainties	Because customer may have different and valuable insight on business and project risks that have been part of the customer expectations but not reflected in real planning	When requirements are being written	Customer risk analysis	Customer representative and functional and project managers, jointly
Lessons learned					
Risk audit	Do a project risk audit following selected projects to evaluate success in anticipating and managing risk	Because insights and documents that can lead to better risk management in the future will be lost unless a risk audit team builds a history of the project, how risk decisions were made and how effective risk management was	At project close-out	Risk audit report to project manager	Project management office, audit staff, project and functional managers
Lessons learned meeting and report	Prepare and communicate short report on what project team members and customers learned in the project that would reduce risks in a similar future project	Because the best lessons and insights are going to be lost unless someone facilitates a lessons learned session and report	At close-out	Lessons learned report, referencing systems, decisions, risk, outcomes but no names	Project manager

Bibliography

Barkley, B. T., and James Saylor, *Customer-Driven Project Management: Building Quality into Project Processes,* 2d ed., McGraw-Hill, New York, 2001.

Bennatan, E. M., *On Time and Within Budget: Software Project Management Practices and Techniques,* Wiley, 2000.

Burgelman, R., Modesto Maidique, and Steven Wheelwright, *Strategic Management of Technology and Innovation,* 3d ed., McGraw-Hill, New York, 2001.

Buttrick, R., *The Interactive Project Workout*, Prentice Hall, 2000.

Harrington, H. J., Daryl R. Conner, and Nicholas Horney, *Project Change Management: Applying Change to Improvement Projects*, McGraw-Hill, New York, 2000.

Keane, *Inc. Productivity Management: Keane's Project Management Approach for Systems Development,* 2d ed., Keane, Inc, 1995.

Kendall, G., and Steven Rollins, *Advanced Project Portfolio Management and The PMO,* J. Ross Publishing, 2003.

Meredith, J., and Samuel Mantel, Jr., *Project Management: A Managerial Approach*, 5th ed., Wiley, 2003.

Project Management Institute, *A Guide to the Project Management Body of Knowledge,* Project Management Institute, 2000.

Project Management Institute, *Practice Standard for Work Breakdown Structures*, Project Management Institute, 2001.

Royer, P. S., *Project Risk Management: A Proactive Approach*, Management Concepts, 2002.

Shtub, A., Jonathan Bard, and Shlomo Glorberson, *Project Management: Engineering, Technology, and Implementation*, Prentice Hall, 1994.

Wideman, R. M., *Project and Program Risk Management: A Guide to Managing Project Risks and Opportunities*, Project Management Institute, 1992.

Index

Page numbers followed by *f* or *t* indicate figures and tables, respectively.